INTRODUCTION

This book is one of a series specially devised to aid the busy professional dealer in his everyday trading. It will also prove to be of great value to all collectors and those with goods to sell, for it is crammed with illustrations, brief descriptions and valuations of hundreds of antiques.

Every effort has been made to ensure that each specialised volume contains the widest possible variety of goods in its particular category though the greatest emphasis is placed on the middle bracket of trade goods rather than on those once-in-a-lifetime museum pieces whose values are of academic rather than practical interest to the vast majority of dealers and collectors.

This policy has been followed as a direct consequence of requests from dealers who sensibly realise that, no matter how comprehensive their knowledge, there is always a need for reliable, up-to-date reference works for identification and valuation purposes.

When using your Antiques and their Values Book to assess the worth of goods, please bear in mind that it would be impossible to place upon any item a precise value which would hold good under all circumstances. No antique has an exactly calculable value; its price is always the result of a compromise reached between buyer and seller, and questions of condition, local demand and the business acumen of the parties involved in a sale are all factors which affect the assessment of an object's 'worth' in terms of hard cash.

In the final analysis, however, such factors cancel out when large numbers of sales are taken into account by an experienced valuer, and it is possible to arrive at a surprisingly accurate assessment of current values of antiques; an assessment which may be taken confidently to be a fair indication of the worth of an object and which provides a reliable basis for negotiation.

Throughout this book, objects are grouped under category headings and, to expedite reference, they progress in price order within their own categories. Where the description states 'one of a pair' the value given is that for the pair sold as such.

The publishers wish to express their sincere thanks to the following for their kind help and assistance in the production of this volume:

JANICE MONCRIEFF
NICOLA PARK
CARMEN MILIVOYEVICH
ELAINE HARLAND
MAY MUTCH
MARGOT RUTHERFORD
JENNIFER KNOX
SHONA BROWN
KAREN KILGOUR

Printed by Apollo Press, Worthing, Sussex, England.
Bound by R. J. Acford, Chichester, Sussex, England.

CONTENTS

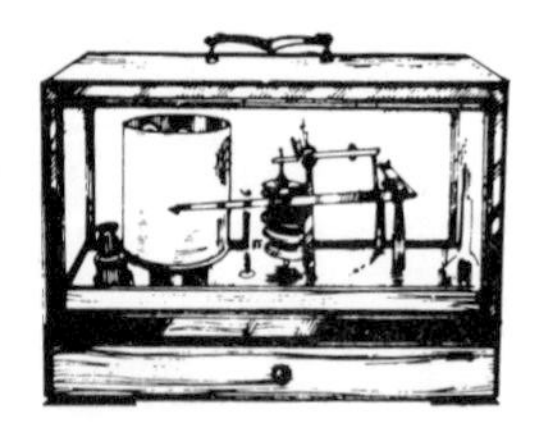

Early 20th century oak barograph by Lennie. $170 £75

Barograph in oak case with drawer, 15in. wide. $200 £90

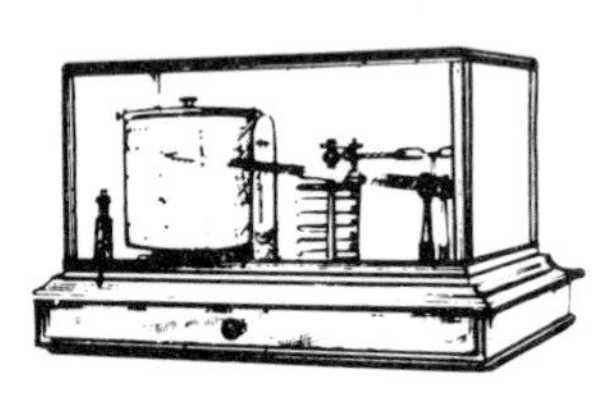

A barograph by Whyte Thomson & Co., Glasgow, in a mahogany and glazed case. $250 £110

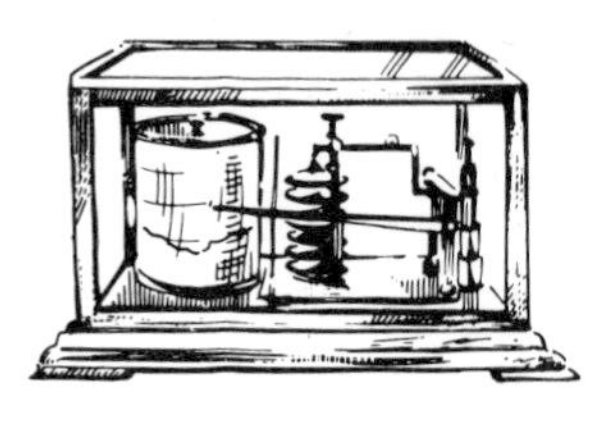

Late 19th century plated and brass barograph by Short & Mason. $280 £125

A fine 19th century clock and barograph by J. L. Cassartelli. $550 £245

A Mariner's clock in brass, plate with a barograph and compass, engraved Thos. Russell & Son, Paris. $1,080 £480

ADMIRAL FITZROY

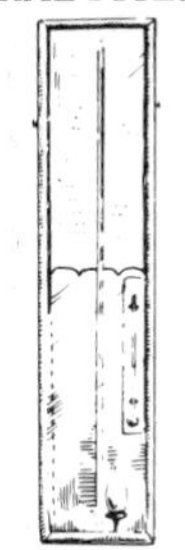

Victorian Admiral Fitzroy barometer in glazed mahogany case, 35in. long. $180 £80

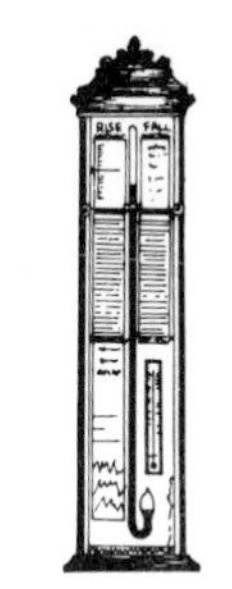

Victorian Admiral Fitzroy barometer in an oak case. $190 £85

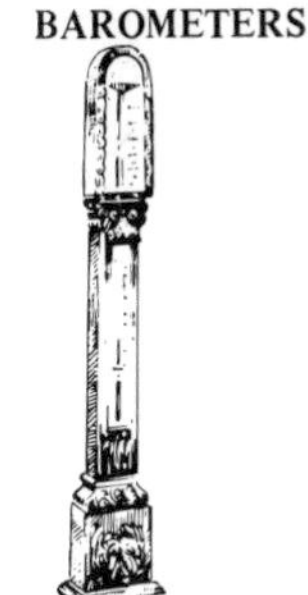

Admiral Fitzroy barometer by Negretti and Zambra in carved oak case. $820 £405

ANEROID

Victorian wall barometer in an oak case. $45 £20

A circular barometer in brass case on ebonised stand. $80 £35

19th century brass cased aneroid barometer set in crossed oars. $160 £70

ANGLE

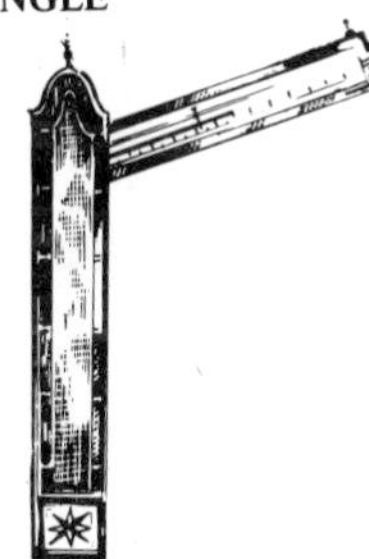

18th century angle barometer in case of banded and inlaid mahogany, inscribed David Fosell, Hinckly. $2,475 £1,100

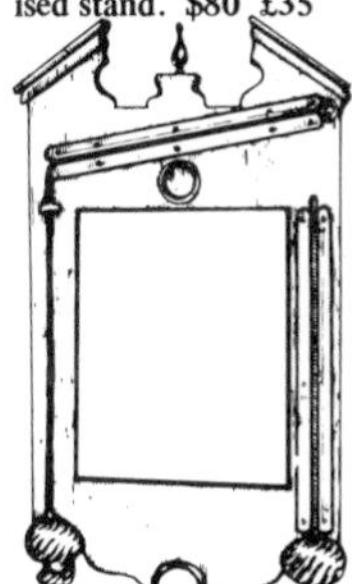

George II mahogany angle barometer and thermometer in the form of a wall mirror, 102.cm. high. $3,375 £1,500

Rare angle barometer in walnut case by Charles Orme, 1736, Ashby-de-la Zouch. $7,080 £3,000

BAROMETERS
BANJO/BROKEN ARCH

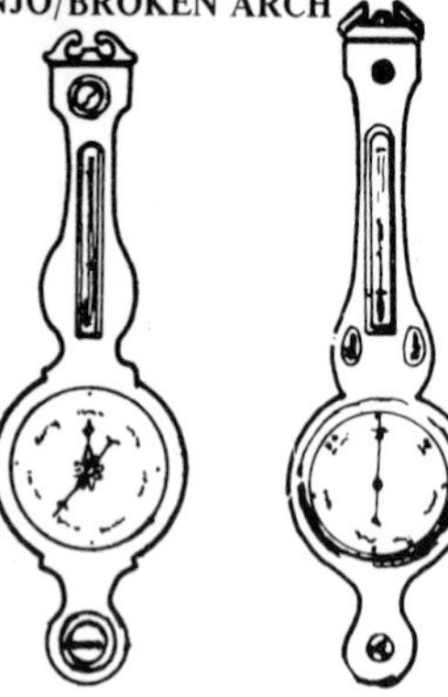

Victorian banjo barometer by Lione & Samalvico of Holborn. $180 £80

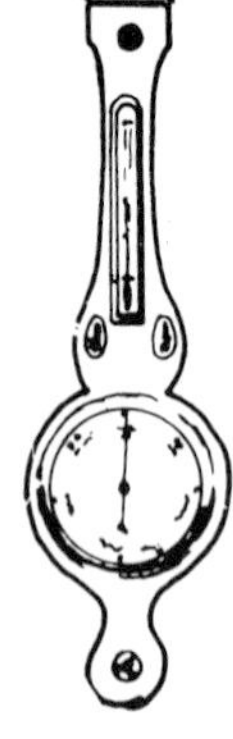

Early 19th century wheel barometer by G. Balsgry, London, in an inlaid mahogany case. $235 £105

Mahogany inlaid wheel barometer and thermometer by Gulletti, Glasgow. $235 £105

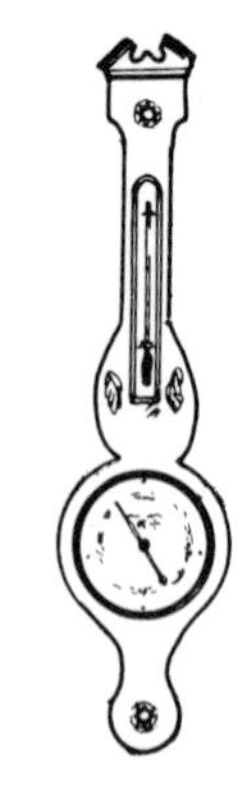

Mahogany inlaid wheel barometer and thermometer of Sheraton design, by A. Riva & Co., Glasgow. $250 £110

A 19th century mahogany inlaid wheel barometer and thermometer. $270 £120

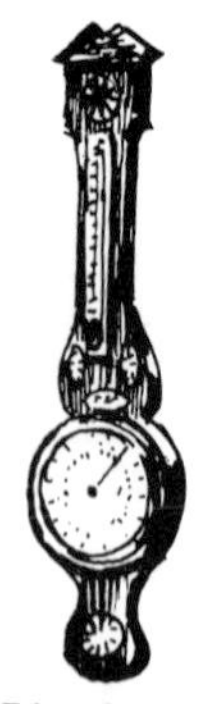

Edwardian banjo barometer and thermometer by C.S.S.A. Ltd. of London. $270 £120

George III inlaid mahogany barometer by M. & E. Levin of London. $305 £135

George III style rosewood barometer by Ciceri Pini & Co., Edinburgh. $325 £140

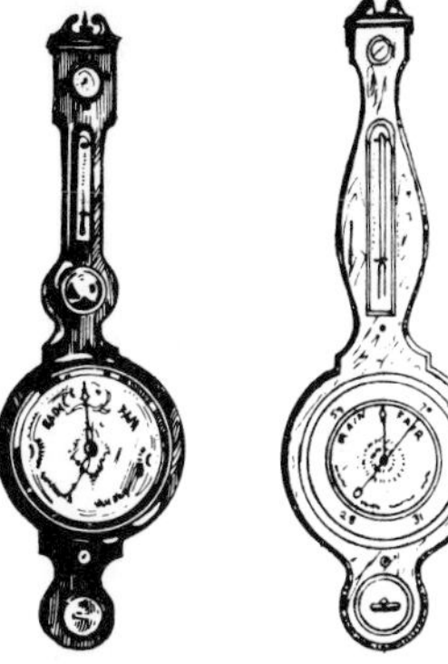

Mahogany inlaid wheel barometer and thermometer, inscribed J. Fagliotti, Clerkenwell. $325 £145

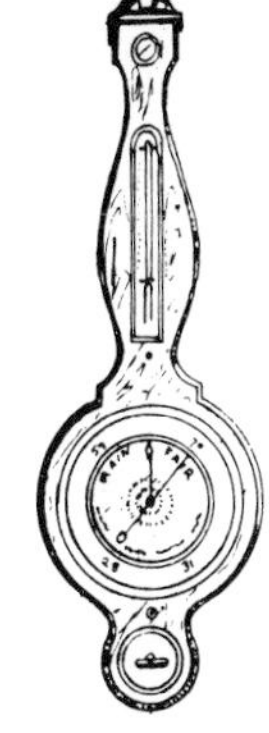

Early 19th century mahogany scroll top barometer with boxwood string inlay. $325 £145

Mahogany wheel barometer with swan neck pediment, ebony and boxwood stringing, 39in. high, circa 1800. $340 £150

A small mahogany wheel barometer and thermometer by Dawson, Haddington, 3ft.2in. high. $350 £155

Late 18th century mahogany cased barometer with satinwood inlay. $350 £155

Early 19th century barometer and thermometer by Taylor & Son, London, 38in. long. $315 £155

George III style mahogany barometer by B. Corji, Aberdeen. $390 £170

Early 19th century barometer by C. A. Canti & Son, London, 38in. high. $440 £200

Mahogany wheel barometer with shell inlay by A. Pagani, Nottingham, circa 1810.
$450 £200

Sheraton wheel barometer by A. Benzoni, circa 1800. $450 £200

19th century barometer by F. Amadio, London, 38in. high. $460 £210

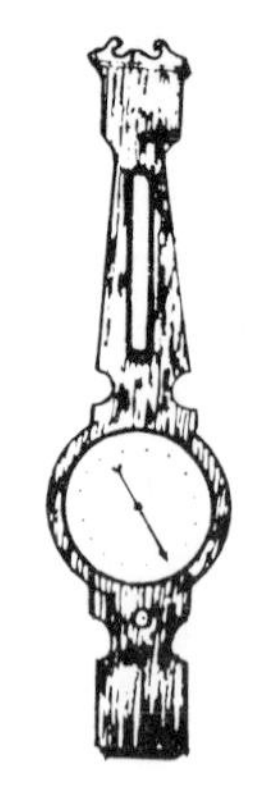

A Regency period rosewood barometer by Myers of Nottingham, circa 1825.
$485 £215

George III mahogany cased barometer by Tara, Louth.
$495 £220

Sheraton period barometer, dial inscribed M. Salamon, Oxford, 38in. high. $485 £220

Wheel barometer by M. Barnuk, in faded mahogany case, 39in. high, circa 1795.
$510 £225

George III mahogany cased wheel barometer by C. Gerlatti, Glasgow.
$540 £240

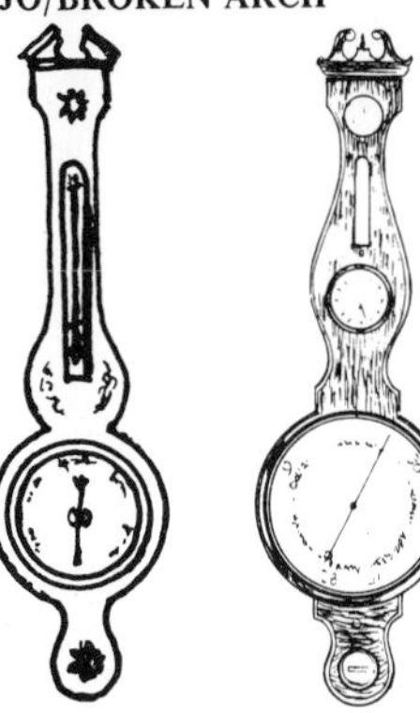

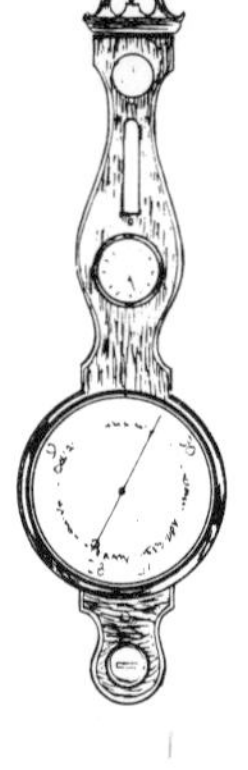

Mahogany 8in. wheel barometer, circa 1790, engraved B. Corty, Glasgow. $540 £240

Early 19th century barometer in a mahogany case complete with eight-day timepiece. $565 £250

A Sheraton wheel barometer, with medallion, conch motifs and stringing, by M. Salla, Preston, 1810. $565 £250

Mid 19th century walnut veneered banjo wheel barometer, signed Stopani, Aberdeen, 95cm. high. $590 £260

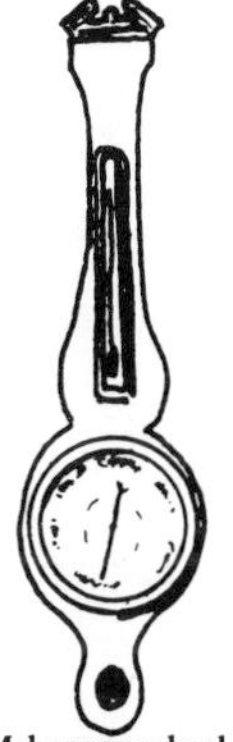

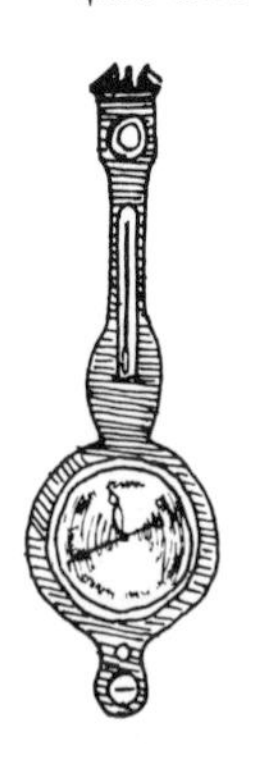

Mahogany cased 8in. diam. wheel barometer, circa 1820, 3ft.2½in. long. $585 £260

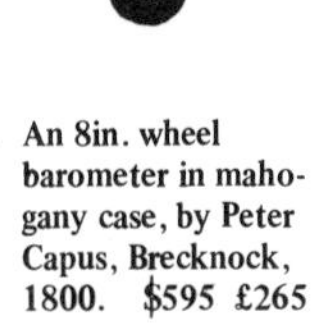

An 8in. wheel barometer in mahogany case, by Peter Capus, Brecknock, 1800. $595 £265

Mahogany wheel barometer by N. Barnuka, Bury, inlaid with ebony and holly paterae, circa 1795. $620 £275

Mahogany cased wheel or banjo barometer, circa 1820, by G. Balsary, 39½in. high. $660 £285

A finely figured mahogany barometer, with hygrometer at the top and spirit level at the bottom, 10in. diam., by I. Sordelli, London, 1800. $790 £350

Rosewood wheel barometer by Emilio Zuccani of London, circa 1840, 44in. high. $835 £370

George III barometer with hygrometer, thermometer and clock. $845 £375

A barometer in a satinwood case, by I. Vecchio, Nottingham, circa 1800. $880 £390

Late Georgian mahogany wheel clock barometer, signed Leonardo Camberwell No. 460, 46in. high. $900 £400

Rosewood wheel barometer with silvered dial, by J. Somalvico & Son, circa 1820, 38in. high. $1,125 £500

Mahogany wheel clock barometer by Pastorelli, London, 47in. high. $1,300 £580

Large clock wheel barometer by I. W. Hancock, Yeovil, 48in. high. $1,590 £710

19th century inlaid mahogany wheel barometer and thermometer by J. Gilmour, Glasgow, 94cm. high. $170 £75

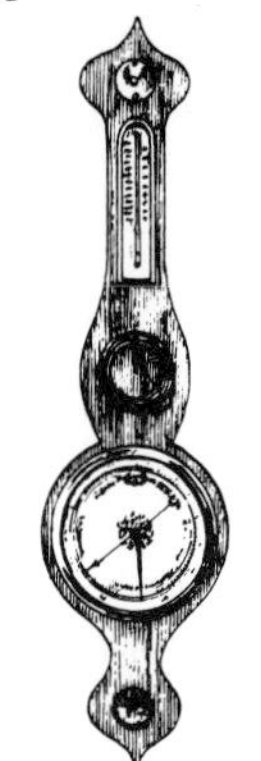

19th century rosewood cased wheel barometer and thermometer. $215 £95

Wheel barometer by J. T. Jeffs of Luton, 38in. high, circa 1850. $305 £135

Rosewood banjo barometer with white enamel dial, circa 1850, 40in. high. $400 £180

Rosewood cased wheel barometer, circa 1840, by Callaghan, Preston, 40½in. high. $420 £185

Large mahogany wheel barometer by Snow, Ripon, circa 1850, with a 10in. dial. $440 £195

Large mahogany banjo barometer, 40in. high, maker's name F. Harrison, Hexham, circa 1830. $450 £200

A fine 19th century rosewood case barometer inlaid with mother-of-pearl. $820 £365

BAROMETERS
BANJO/SHAPED TOP

Victorian oak cased banjo barometer and thermometer. $100 £45

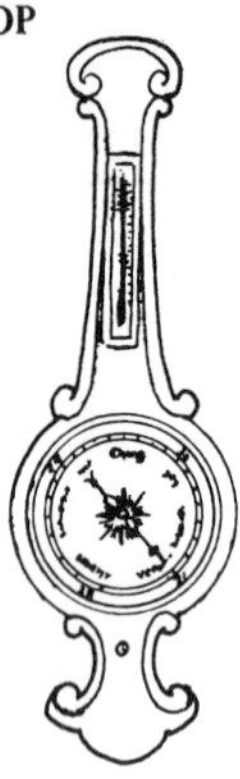

Victorian banjo barometer and thermometer in a rosewood case. $125 £55

Late Victorian barometer and thermometer by Swindon & Sons, Birmingham, in a carved oak case. $160 £70

A rosewood wheel barometer by Field & Sons. $170 £75

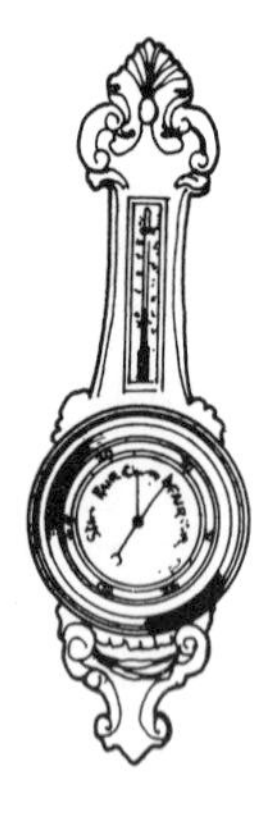

Victorian carved oak aneroid barometer and thermometer. $180 £80

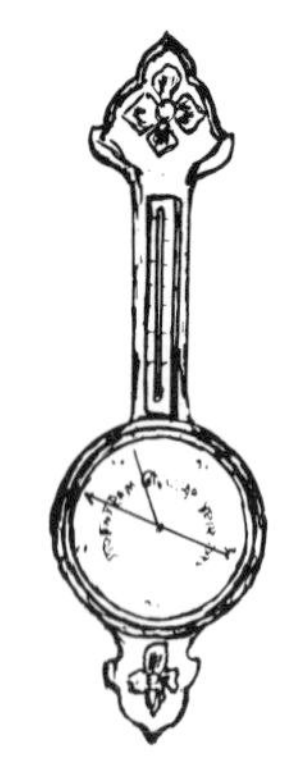

19th century mahogany cased barometer inlaid with boxwood. $250 £110

Victorian rosewood cased barometer by K. Leyser & Co., London, inlaid with mother-of-pearl, 39in. high.

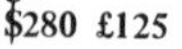

$280 £125

Rosewood wheel barometer with 10in. dial, circa 1860. $305 £135

A 19th century rosewood and mother-of-pearl inlaid banjo barometer and thermometer, by Green, Glasgow.
$340 £150

Late 18th century finely figured and inlaid Sheraton wheel barometer. $565 £250

Late 18th century mahogany wheel barometer with silvered dial, circa 1790, 37in. high.
$675 £300

George III mahogany banjo barometer, 38in. high.
$900 £400

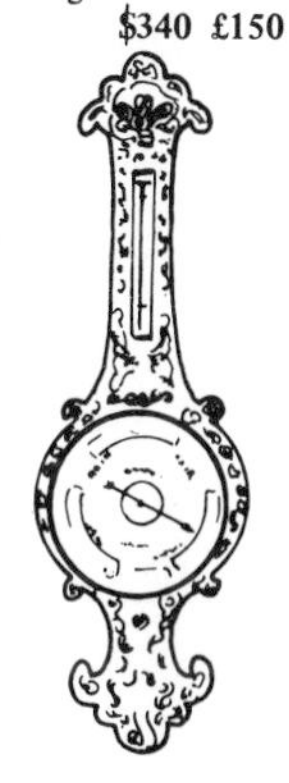

Rosewood and mother-of-pearl inlaid barometer, 19th century.
$990 £440

Mother-of-pearl inlaid rosewood barometer, circa 1850, 44¾in. high.
$1,070 £475

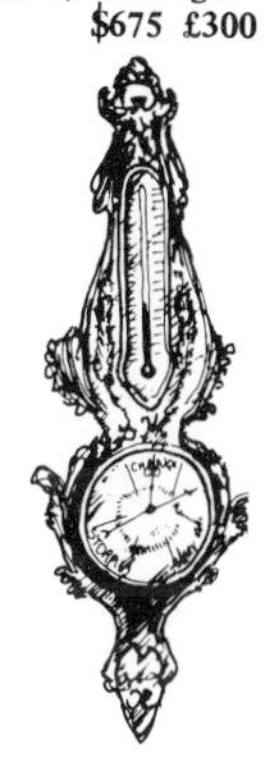

A fine late 18th century French banjo barometer decorated with gilt and ormolu, 44in. high. $1,125 £500

A fine late 18th century French barometer with ormolu decoration.
$1,690 £750

A small 19th century timepiece and barometer in brass horseshoe pattern case, 6in. wide. $90 £40

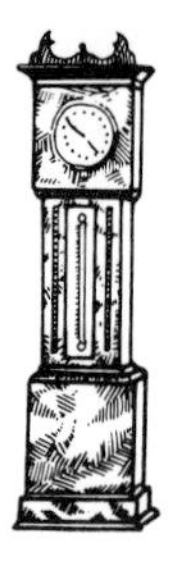

Miniature oak grandfather clock and barometer, 16in. high. $135 £60

19th century German barometer clock with an oak case. $295 £130

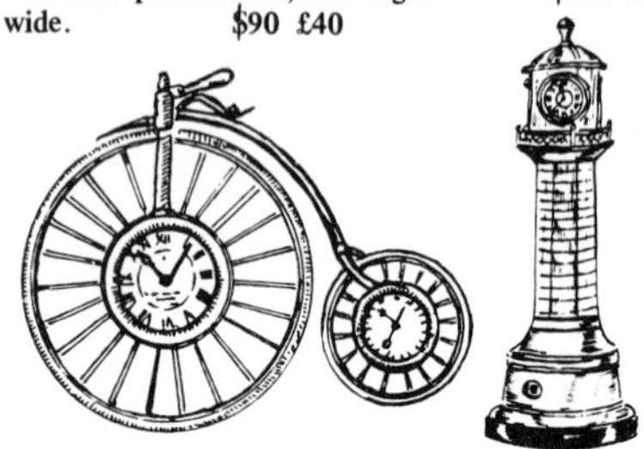

Pennyfarthing timepiece and aneroid barometer. $315 £140

English brass lighthouse combining clock and barometer. $450 £200

An oval timepiece below which is an aneroid barometer, with thermometers at either side, 6¾in. high. $620 £275

An unusual combination timepiece with aneroid barometer combined in one case, by Richard et Cie, Paris, 5in. high. $620 £275

Late 19th century combined clock, thermometer, barometer and barograph, 2ft.3in. wide. $675 £300

An astronomical calendar clock in rouge case, by Lister & Son of Newcastle. $1,125 £500

Early 18th century walnut pillar barometer, 3ft.3in. high.
$1,745 £775

Walnut portable pillar barometer by Daniel Quare, circa 1700, 3ft.3in. high.
$1,755 £780

Early 18th century portable barometer in a walnut case.
$2,475 £1,100

Very rare George II column barometer, 2ft.11in. high.
$2,700 £1,200

Carved walnut portable pillar barometer by Daniel Quare, London, circa 1710, 3ft.2in. high. $3,300 £1,500

Rare late 17th century ebonised column barometer by Thos. Tompion, 3ft.5in. high.
$5,625 £2,500

A rare walnut pillar barometer by Daniel Quare, 3ft. high.
$7,315 £3,250

18th century walnut cased stick barometer by D. Quare, 39in. high.
$9,560 £4,250

BAROMETERS SHAPED CASE

Late 19th century barometer and thermometer in mahogany inlaid case. $90 £40

Oak barometer and timepiece, 1880's, 44in. high. $440 £200

George III barometer by Bregazzi of Nottingham, in a mahogany case. $730 £325

George II walnut barometer with circular silvered register plate, 3ft.8in. high. $1,575 £700

Louis XV barometer in boulle case, complete with wall bracket. $1,935 £860

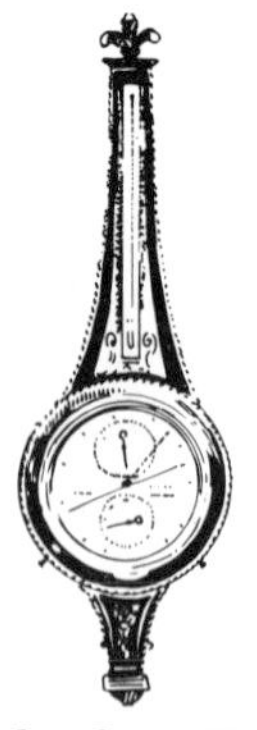

Late George III mahogany wheel barometer, signed Jno. Russell, Falkirk, 120cm. high. $8,280 £3,600

Mahogany wheel barometer by J. Vulliamy, in a case by John Bradburn, 4ft.2in. high. $17,550 £7,800

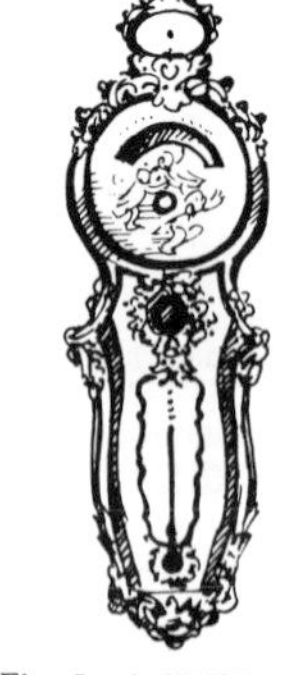

Fine Louis XV kingwood and tulipwood cartel barometer, 42in. high, with painted face. $37,400 £17,000

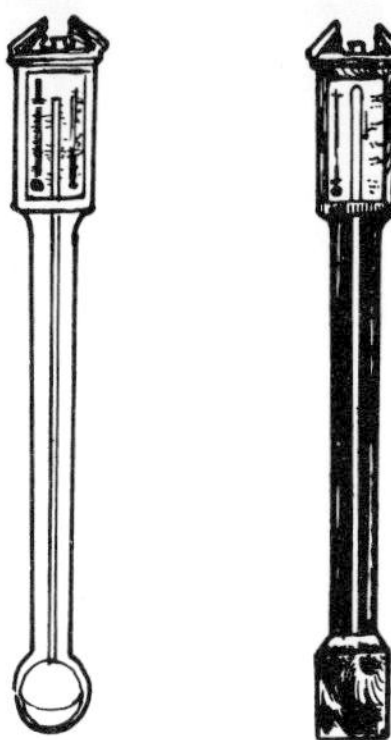

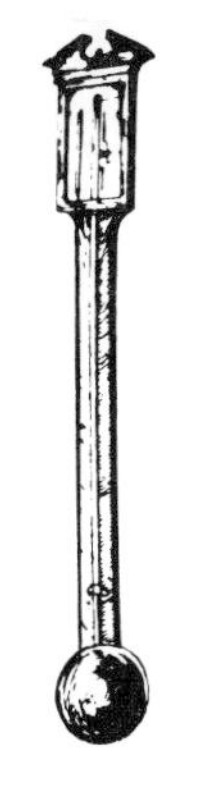

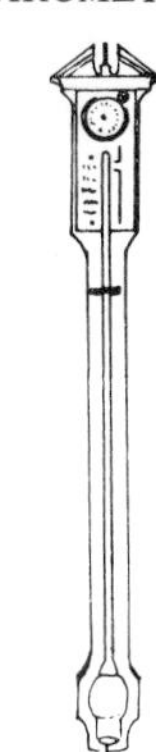

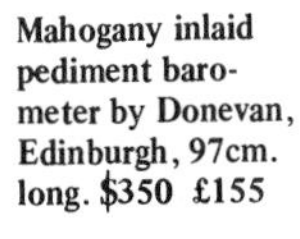

Mahogany inlaid pediment barometer by Donevan, Edinburgh, 97cm. long. $350 £155

Mahogany inlaid pediment barometer by Champion of Glasgow, 96cm. long. $495 £220

George III mahogany inlaid pediment mercury barometer by O. Garof & Co., Edinburgh, 3ft.1in. long. $565 £250

Late Georgian stick barometer by Watkins & Smith, London, in a mahogany case. $585 £260

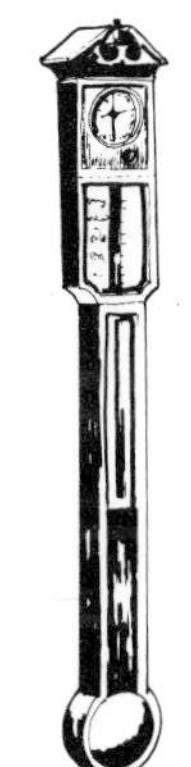

A mahogany stick barometer by J. Gatty, with a humidity dial. $700 £310

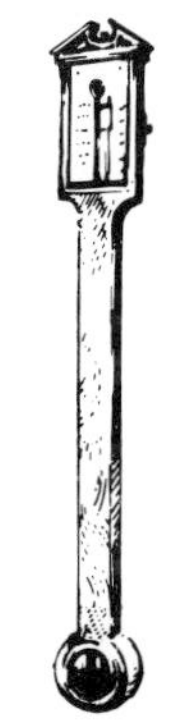

George III mahogany stick barometer by George Adams of Fleet Street. $830 £410

George II mahogany stick barometer by Carls. Aiano, 3ft.4in. high. $1,125 £500

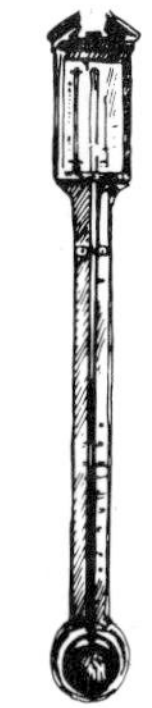

George III mahogany stick barometer by D. Fagoli, 38in. high. $1,175 £520

BAROMETERS
STICK/BROKEN ARCH

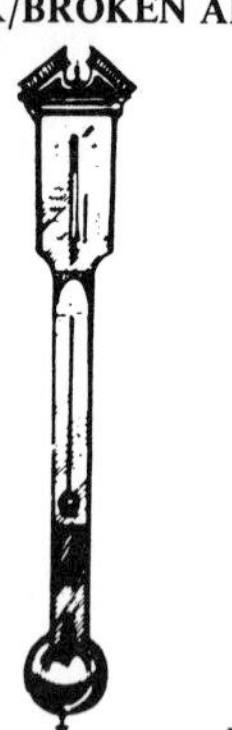

18th century mahogany stick barometer by Pyefinch, London, 3ft.3in. long.
$1,240 £550

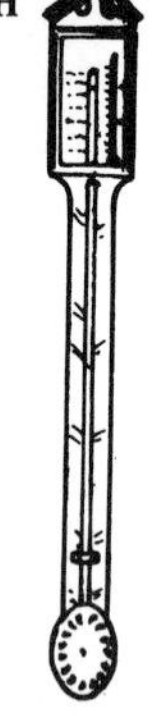

Late George III yewwood stick barometer signed Jas. Corte, Glasgow, 3ft. 2½in. high.
$1,465 £650

Inlaid mahogany stick barometer by Negrety, 1850, 44in. long.
$1,630 £725

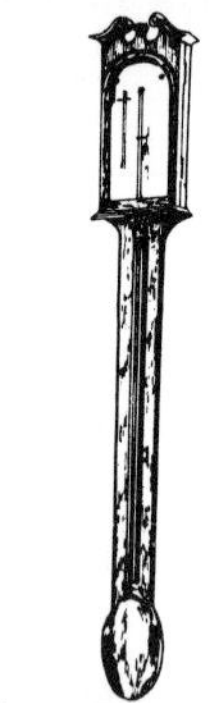

George III mahogany inlaid pediment mercury barometer, by Jn. Russell, Falkirk, 3ft.6in. high.
$3,375 £1,500

STICK/ROUND TOP

19th century mahogany framed pediment barometer by Gardners, Glasgow.
$215 £95

Mahogany pediment barometer by Ramage of Aberdeen. $225 £100

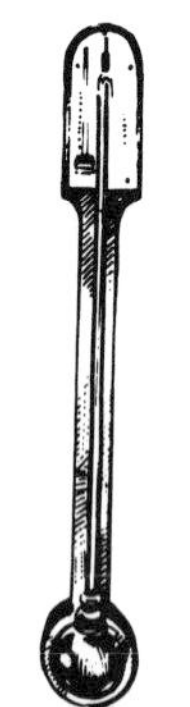

19th century stick barometer with engraved ivory scale in rosewood case, 36in. long.
$360 £160

Rosewood stick barometer by Lancaster & Sons, circa 1830.
$540 £240

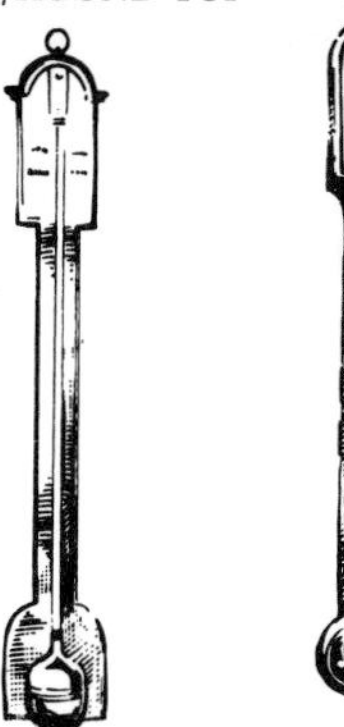

George III mahogany stick barometer, circa 1790, 37in. high.
$555 £245

19th century stick barometer, by J. Pasini, Dorchester, 36in. high.
$585 £260

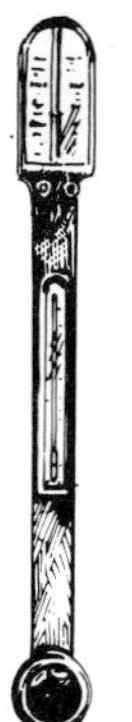

George III rosewood stick barometer by L. Casella & Co. $725 £320

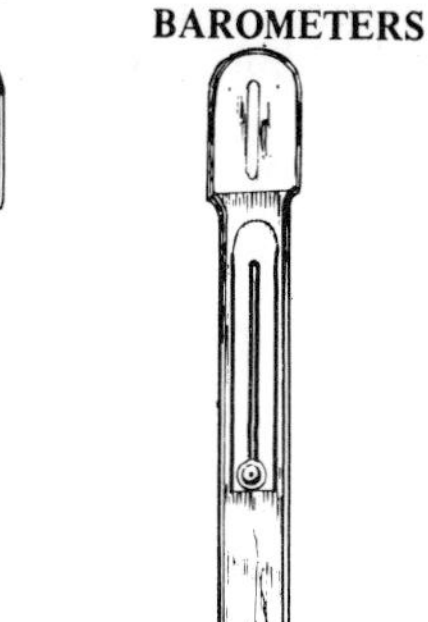

A George IV, mahogany cased, pediment barometer and thermometer with brass mounts, by J. Linnell, London.
$775 £345

A rare Regency ship's barometer, in a fine mahogany case with brass mounts.
$1,170 £520

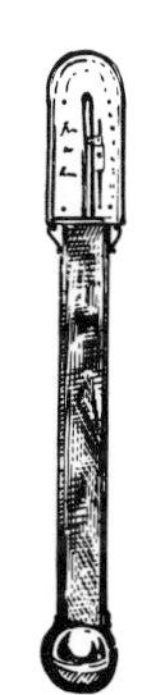

Mid 19th century mahogany stick barometer by B. Martin, London, 90cm. high.
$1,305 £580

18th century mahogany stick barometer by Ed. Nairne, London, 37in. high.
$1,525 £680

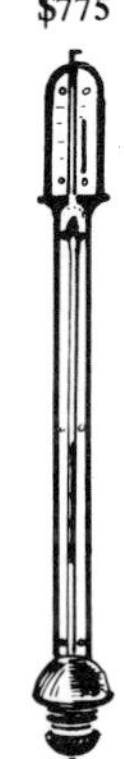

Walnut stick barometer with brass register plate, 2ft. 11½in. high.
$2,250 £1,000

BAROMETERS
STICK/SHAPED TOP

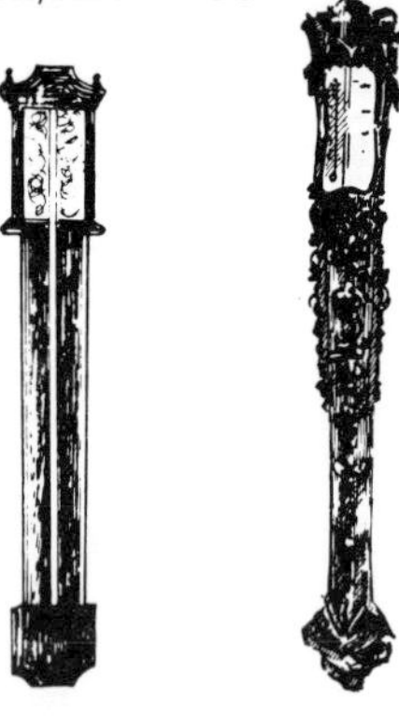

Mahogany cased George III period stick barometer, by Couti. $510 £225

A finely carved 19th century mahogany stick barometer. $735 £335

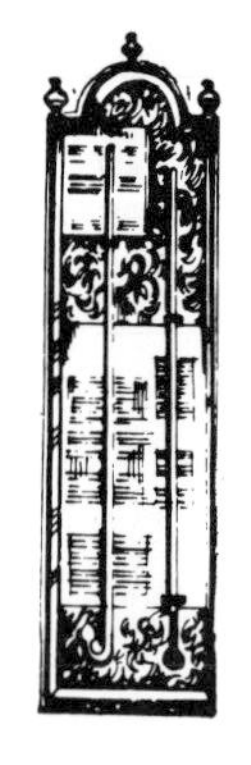

18th century gilt framed barometer and thermometer. $845 £375

Early 17th century barometer in mahogany case. $955 £425

Mahogany stick barometer by Dollond, London, 102cm. high. $990 £440

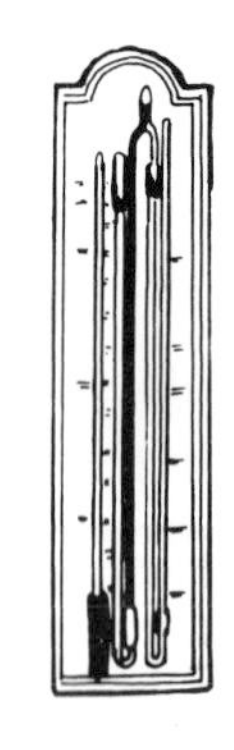

Rare George III double tube barometer, 1ft.11in. high. $1,180 £525

A Georgian mahogany stick barometer with thermometer by Whitehurst & Son, Derby, 3ft.3 3/8in. long. $1,350 £600

Early 18th century walnut stick barometer by Simon Cade of Charing Cross, 37in. long. $1,755 £780

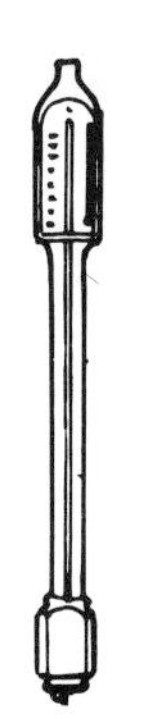

George III mahogany stick barometer, 3ft.5in. high, by Knie, Edinburgh. $1,800 £800

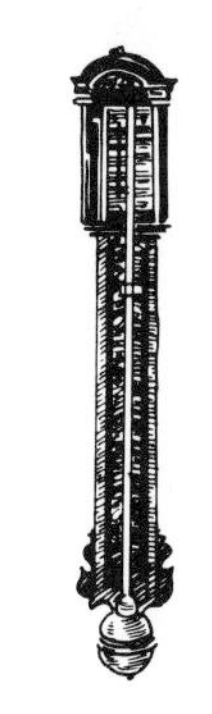

Walnut marquetry barometer, 3ft.2½in. high, with brass register plate. $2,080 £925

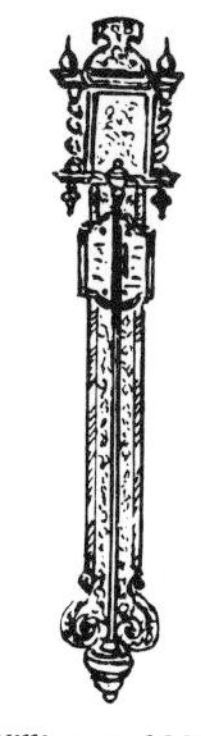

William and Mary walnut marquetry stick barometer, circa 1680, 49in. high. $2,200 £1,000

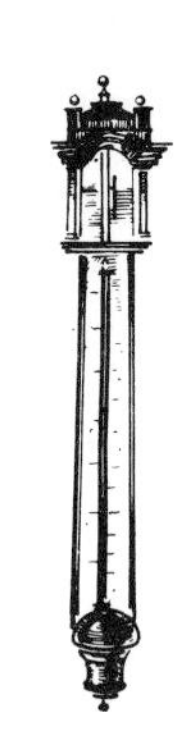

Good George I walnut barometer, circa 1720, 3ft.5½in. high. $2,700 £1,200

Unusual Dutch walnut barometer, mid 18th century, 3ft.11in. high. $2,475 £1,100

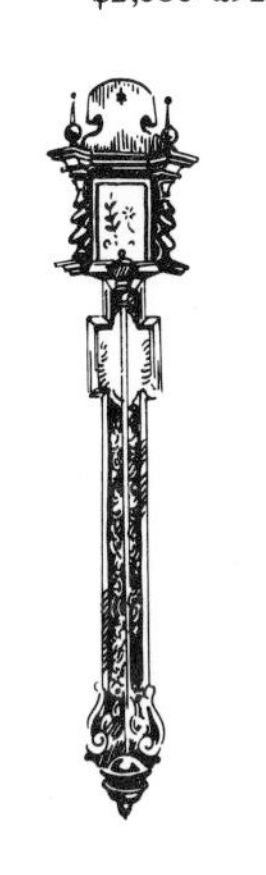

A fine Charles II walnut marquetry stick barometer. $7,650 £3,400

George I barometer by J. N. D. Halifax of Barnsley, 1694-1750. $7,875 £3,500

One of a pair of late Louis XIV Marine barometers attributed to Andre Charles Boulle, 5ft. high. $135,000 £60,000

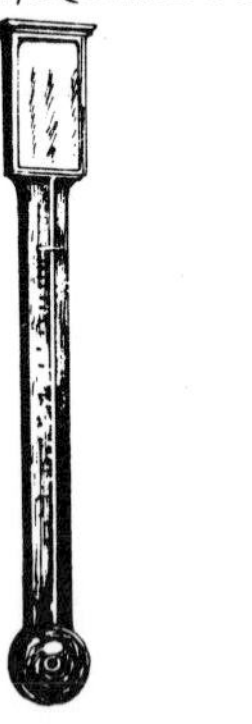

19th century pediment barometer and thermometer by G. Branchi, Edinburgh, 3ft.1in. long. $360 £160

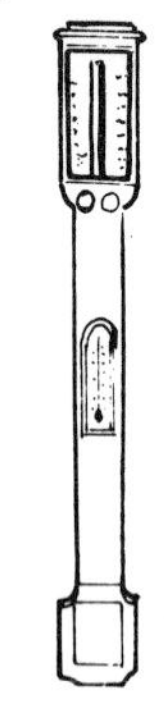

Walnut cased cistern stick barometer by J. Perse, Winchester, circa 1845, 39½in. high. $575 £255

Mahogany stick barometer by Ciceri & Pini, Edinburgh, circa 1855. $675 £300

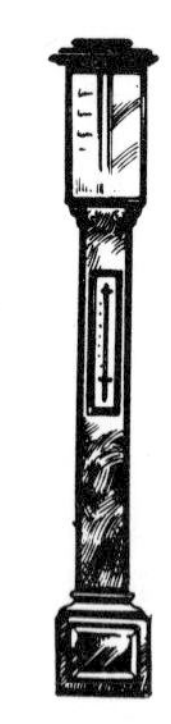

Victorian mahogany stick barometer signed F. W. Clarke, London, with ivory register plates, 38in. high. $635 £340

Early 19th century mahogany stick barometer, signed T. Dunn, Edinburgh, 3ft.1in. high. $1,060 £470

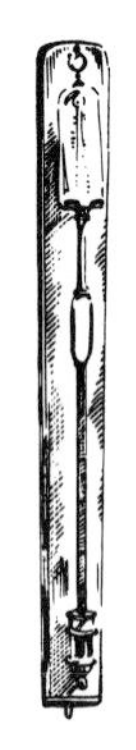

Queen Anne walnut stick barometer with domed pediment. $1,340 £580

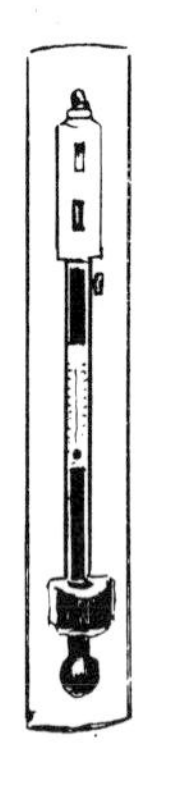

Fortin barometer by J. Hicks, London, mahogany back plate, 50in. high. $1,240 £550

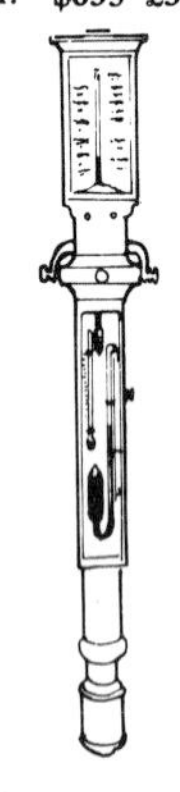

19th century mahogany Marine barometer by J. Morton & Co., Glasgow, 36in. high. $1,755 £780

BRACES

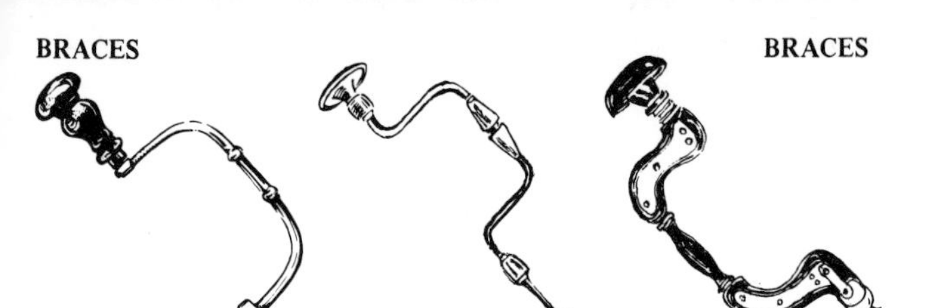

Steel framed wheelbrace with revolving handle in beechwood, 11in. long. $65 £30

Steel Scotch pattern brace by F. Soakes. $108 £50

Beechwood and brass-plated brace by Henry Dixon, Sheffield. $130 £60

CABINETS

Victorian housekeeper's wooden key case for duplicate keys. $55 £25

A laboratory microscope slide cabinet, in white painted pine, the glass door giving access to eighteen drawers, circa 1860. $170 £75

A pine, watchmaker's cabinet, circa 1800, 23in. high, 12in. wide, 7½in. deep. $190 £85

CALIPERS

Engineer's polished steel calipers with curved flat arms, circa 1860. $35 £15

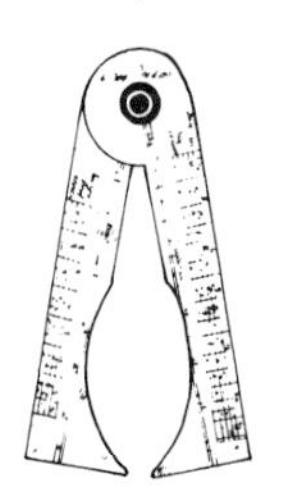

Early 19th century pair of brass Gunners calipers, 176mm. long, in original shaped case. $845 £375

Pair of brass calipers by Culpeper. $1,180 £525

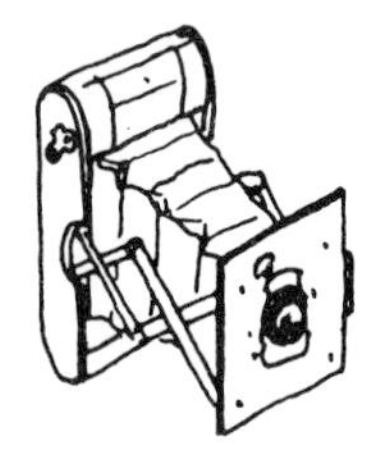

Kodak folding pocket camera. $35 £15

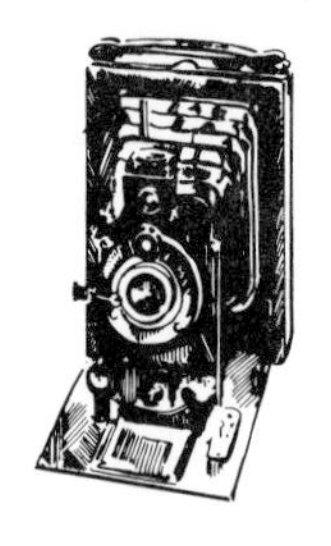

Fine Ica Sirene 135 folding plate camera, 9 x 12cm. $35 £15

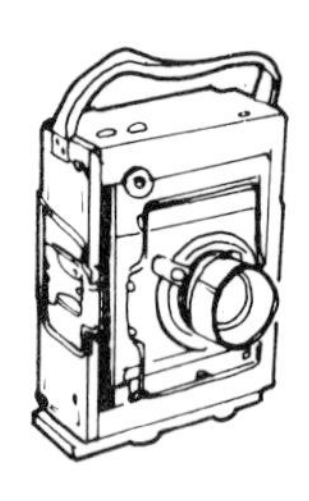

Late 19th century wood and brass plate camera. $45 £20

Fine Ernemann Liliput plate camera, circa 1925, in original box. $70 £30

Ernemann Rolf II folding roll-film camera in original box, circa 1925. $85 £35

German Ernemann Liliput plate camera, circa 1925, with single plate holder. $85 £35

A quarter plate folding 'Sybil' camera, by Newman & Guardia. $90 £40

'Salex' reflex camera. $90 £40

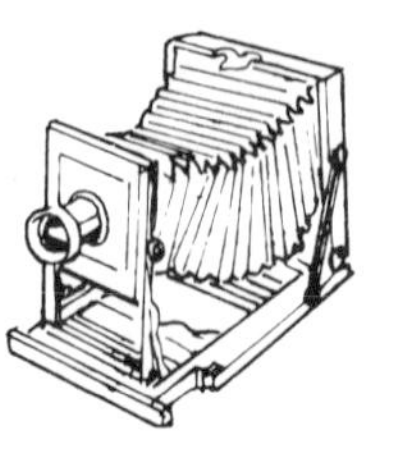

Mahogany brass bellows camera by the Tella Camera Co., circa 1890. $90 £40

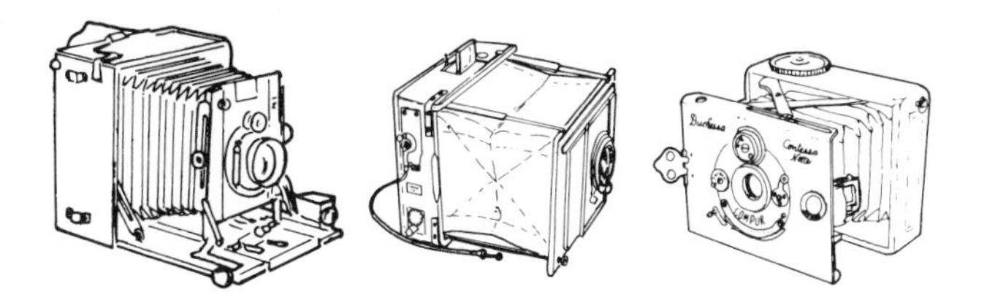

Sanderson hand camera. $100 £45

Goerz 'Anschutz' camera marketed by Pelling & Van Neck, circa 1918. $125 £55

The 'Duchess' miniature plate camera, circa 1920. $125 £55

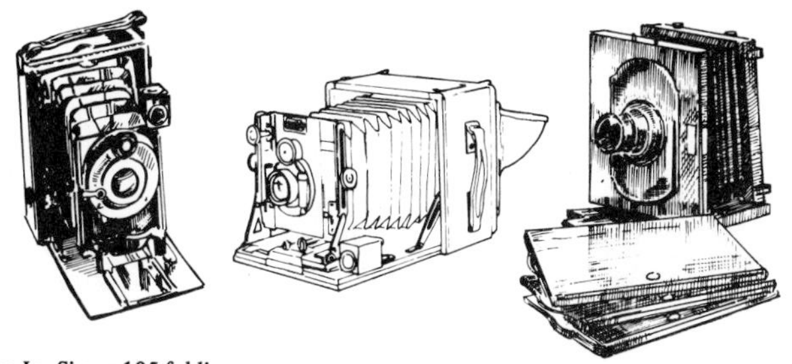

Fine Ica Sirene 105 folding plate camera, 6.5 x 9cm., circa 1925, with instruction booklet, all in original box. $130 £55

Houghton Ltd., 'Sanderson' hand camera, circa 1910. $135 £60

J. Lancaster half-plate 'instantograph' collapsible camera, circa 1905. $145 £65

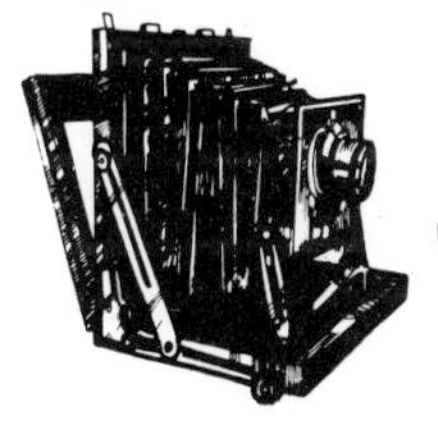

A dry plate camera of mahogany and brass with original lacquer and detachable lens. $170 £75

Dry plate camera by J. Lancaster & Son, Birmingham. $170 £75

A mahogany and brass dry plate camera by J. Lancaster & Son, Birmingham. $170 £75

Quarter-plate camera by J. Lancaster, circa 1893. $180 £80

McKellens double pinion treble patent camera, circa 1890. $180 £80

Late 19th century mahogany and brass half-plate camera, by T. Pickard. $180 £80

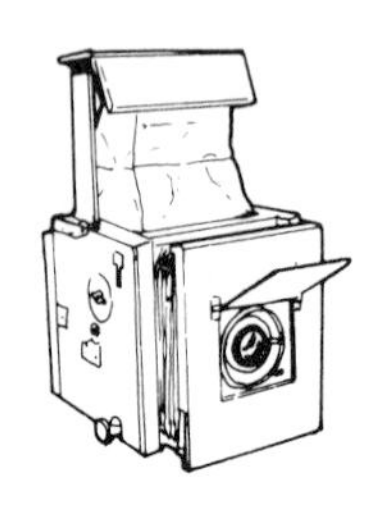

A Newman & Guardia reflex camera, circa 1907. $180 £80

A stand camera, by Lancaster of Birmingham, with revolving disc shutter powered by a rubber band, circa 1895. $180 £80

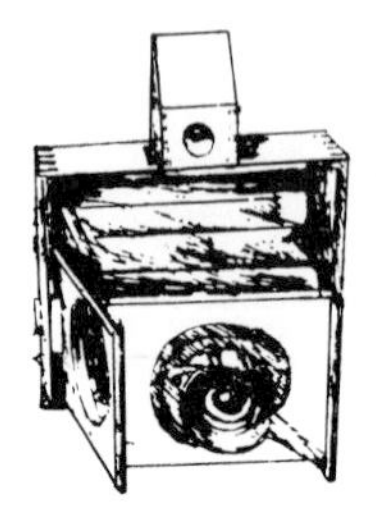

A Shew 'Eclipse' quarter plate, folding camera, circa 1885. $180 £80

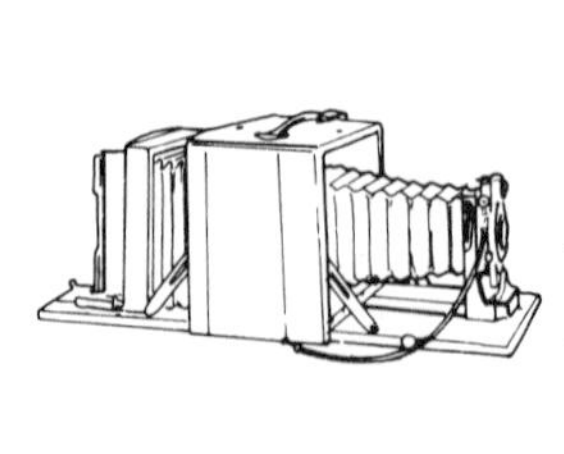

American Folmer & Schwing 'graphic' camera, circa 1920. $190 £85

German Goerz folding reflex Ango camera, circa 1908, 23.5cm. high. $215 £90

A Contax 1 miniature camera by Carl Zeiss, circa 1934. $215 £90

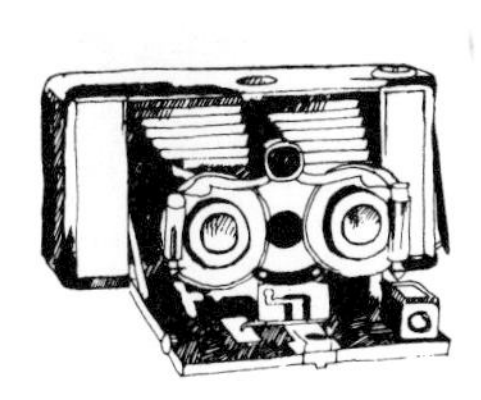

'Stereo Hawkeye' folding camera by Blair Camera Co., model No. 2. $200 £90

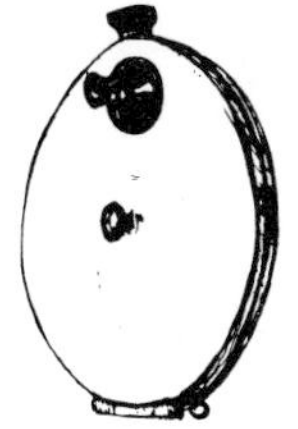

A Stirn secret camera, worn under a coat with the lens projecting through a buttonhole, circa 1885. $215 £95

French Sangor-Shepherd 'The Myrioscope' stereoscopic camera by Gaumont, circa 1900. $230 £95

A rare wet-plate camera by Murray & Heath of Jermyn St., London, 5 x 4in., circa 1865. $215 £95

Rare German Kodak Duo 620 Series II camera, circa 1939. $245 £100

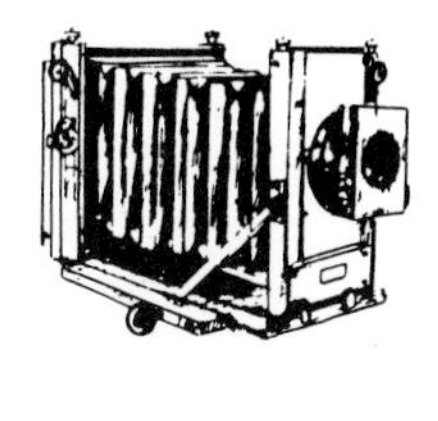

A Sands Hunter 'Imperial' half-plate camera, with boxed roller blind shutter mounted on the front of the lens, circa 1888. $235 £105

Kodak 4a folding camera, negative size 4½ x 6½in. $235 £105

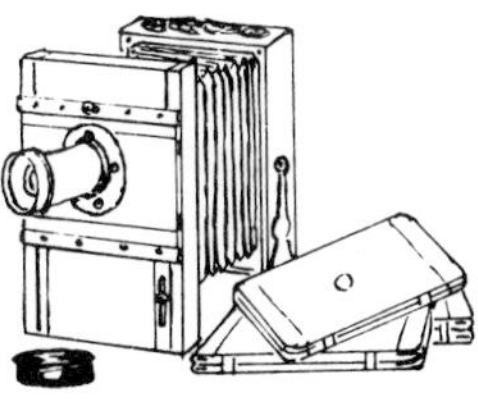

Good tailboard folding field camera, circa 1920, with three double sided plate holders. $235 £105

German Kamera-Werkstatten patent etui 'de luxe' folding camera, circa 1927, 9cm. high. $265 £110

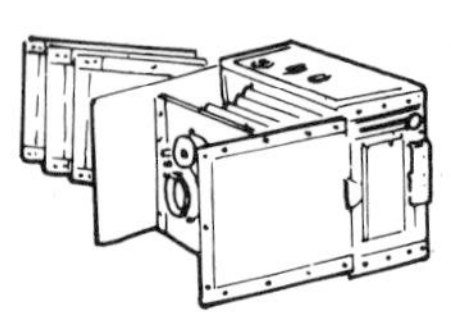

Shew & Co. hand camera, English, circa 1910. $250 £110

London Stereoscopic Co., whole plate camera, circa 1890. $250 £110

A Voigtlander Bess II camera. $250 £110

A Sanderson's Patent 19th century dry plate camera with red leather bellows, the mahogany body covered in black leather. $280 £125

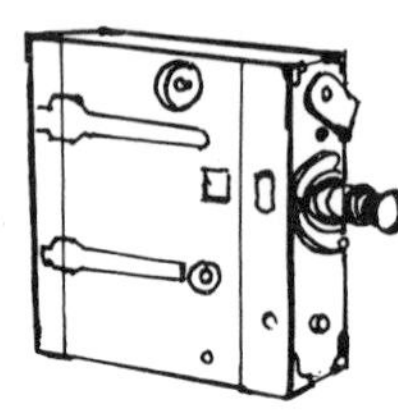

Early 30mm. film camera, circa 1915. $295 £130

19th century mahogany and brass dry plate camera engraved 'Patent Thornton Pickard'. $295 £130

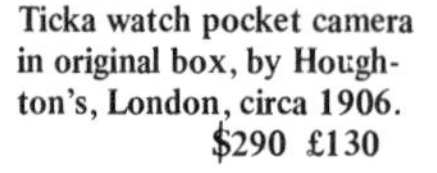

Ticka watch pocket camera in original box, by Houghton's, London, circa 1906. $290 £130

A very fine half-plate 'Royal Ruby' triple extension camera by Thornton Pickard, circa 1912. $295 £130

Leica I 'model A' camera, circa 1930, complete with brown leather case. $340 £140

Austrian Goerz minicord twin lens reflex camera, circa 1951. $360 £150

Fine Sanderson teak wood field camera, 22cm. high, circa 1910, in leather case. $365 £150

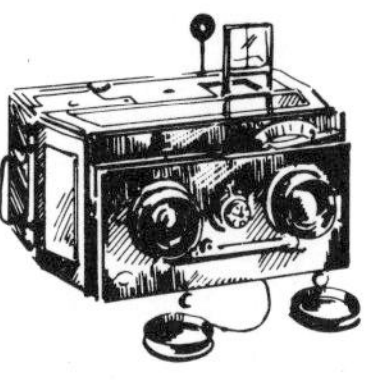

Bellieni stereo jumelle camera, French, 1900, with twin Zeiss Protar lenses. $350 £150

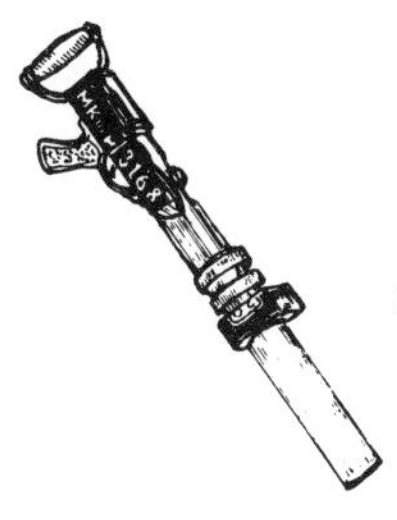

Unusual 'machine-gun' Thornton Pickard camera, circa 1942, 39in. long. $370 £165

A quarter-plate, tropical 'Una' hand-and-stand camera by Sinclair of Haymarket, Edinburgh, in Spanish mahogany and brass, circa 1914. $405 £180

A twin lens reflex 'Pilot' camera, circa 1930. $405 £180

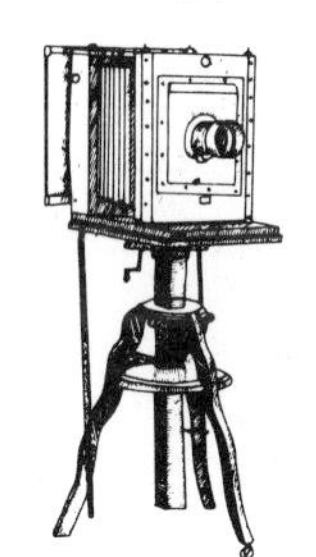

English 15 x 15in. studio camera on tripod stand, circa 1880. $435 £190

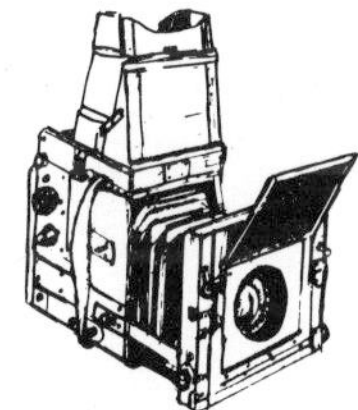

A rare Soho press reflex quarter-plate tropical camera in teak, with red Russian leather bellows and hood, by Kershaw of Leeds. $430 £190

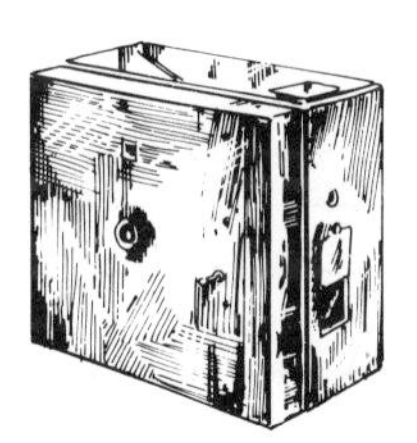

Rare Fallowfield facile 'detective' camera by Miall, circa 1890, 24.7cm. high. $465 £190

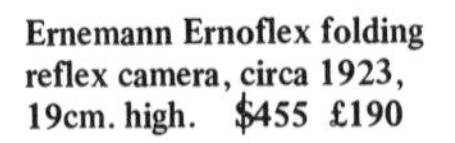

Ernemann Ernoflex folding reflex camera, circa 1923, 19cm. high. $455 £190

German Leica camera, 36cm. long, circa 1950, in Leitz ever-ready case. $460 £190

A press reflex folding tropical 'Minex' camera in teak, brass and tan Russian leather, by Adams, circa 1923. $430 £190

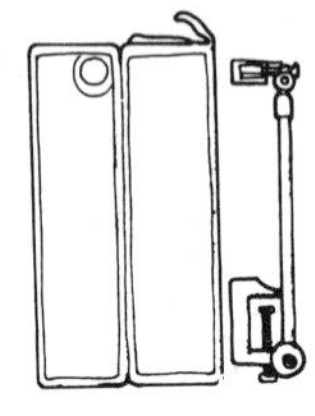

An early 19th century telescopic camera 'Lucida' in original plush lined fitted box. $450 £200

W. Watson & Sons Ltd., full plate tailboard studio camera, circa 1930. $475 £210

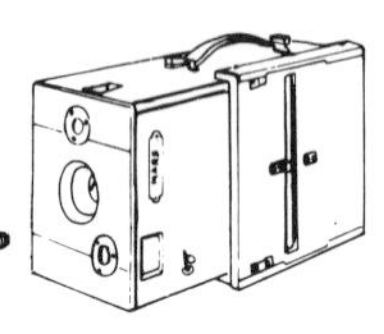

Mars detective camera, 5 x 4½ x 10in., German, circa 1895. $475 £210

A whole plate model of a field camera by Rouch of the Strand, circa 1887. $495 £220

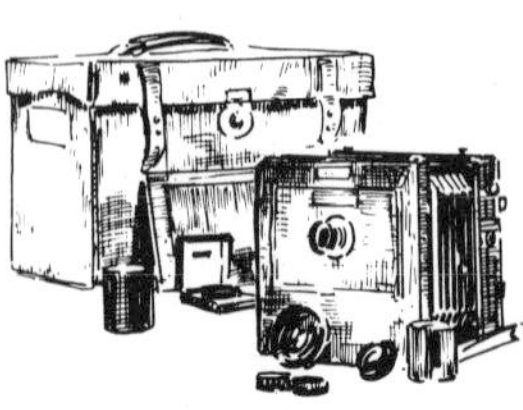

Good Van Neck & Co. tailboard whole plate camera, circa 1895, in original case. $495 £220

Ensign tropical roll-film reflex camera, circa 1925. $495 £220

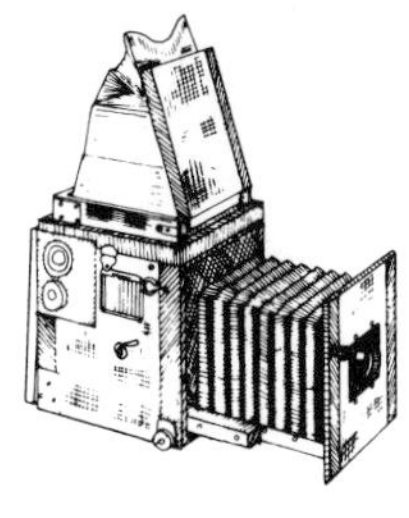

Unusual mahogany reflex camera, circa 1920's, 9 x 12cm. $675 £300

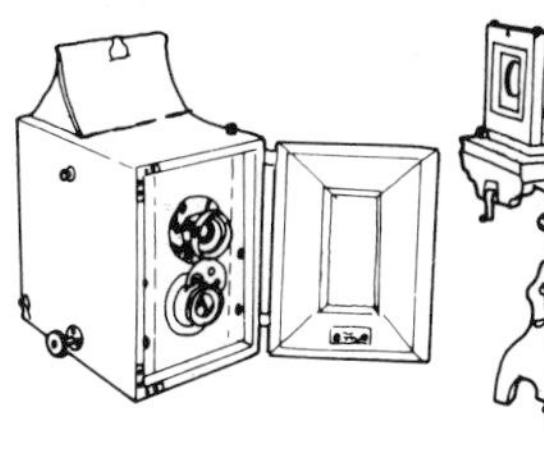

Twin lens reflex, quarter-plate camera by Watson of Holborn, circa 1890. $675 £300

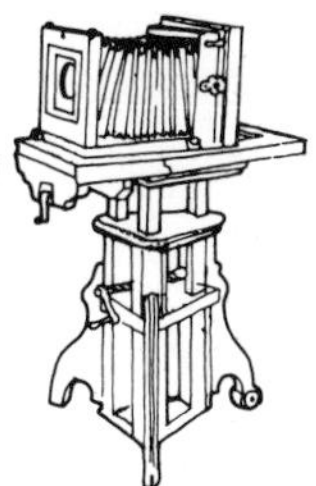

A whole plate studio camera and stand, circa 1900. $700 £310

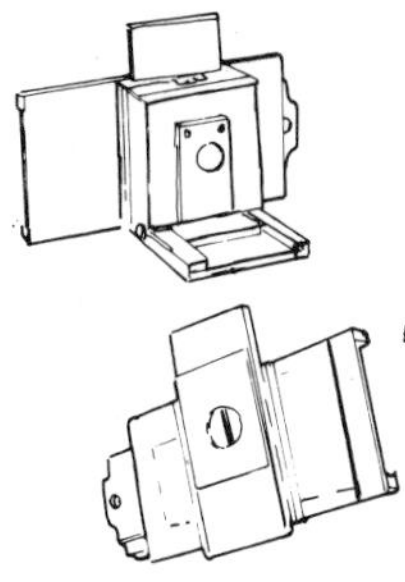

Rare multiple image camera by Billcliff of Manchester, circa 1880. $730 £325

A sliding box wet plate camera by Ottewill of Islington, in mahogany and brass case, 10 x 8in., circa 1860. $790 £350

Sanderson's quarter-plate tropical hand and stand camera, English, circa 1912. $900 £400

Swiss compass camera, 2½ x 2¼ x 1¼in., circa 1940. $900 £400

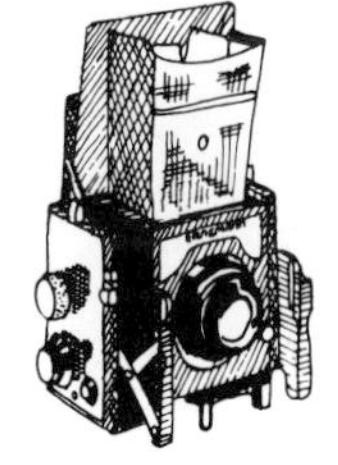

Rare Ernemann Ernoflex folding reflex camera, German, circa 1925. $900 £400

Golden Ernemann high speed press camera, circa 1925. $970 £430

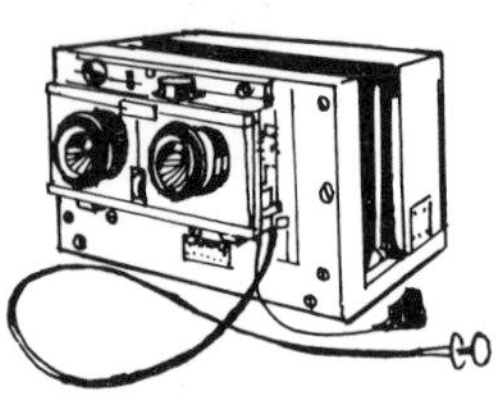

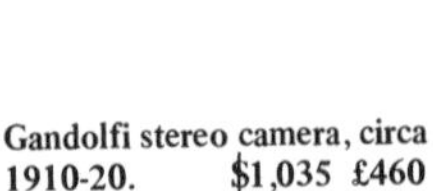

Gandolfi stereo camera, circa 1910-20. $1,035 £460

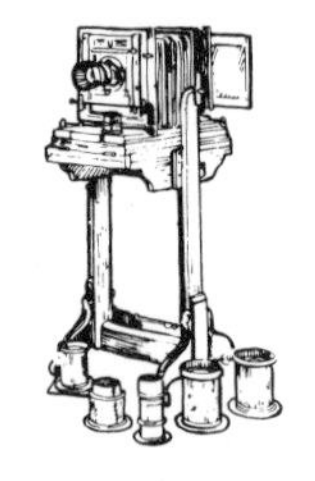

Good German full-plate studio camera by Staeble, Munich, with lenses, circa 1880. $980 £450

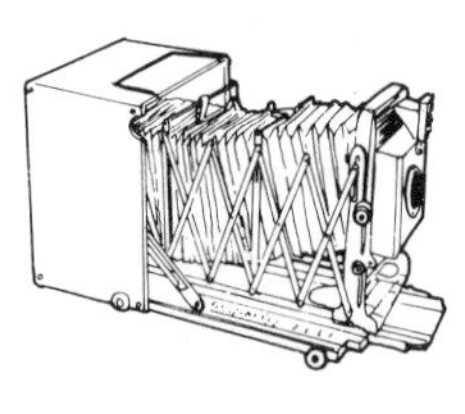

'Trellis' camera by Newman & Guardia, circa 1920, plate size 5 x 4in. $1,070 £475

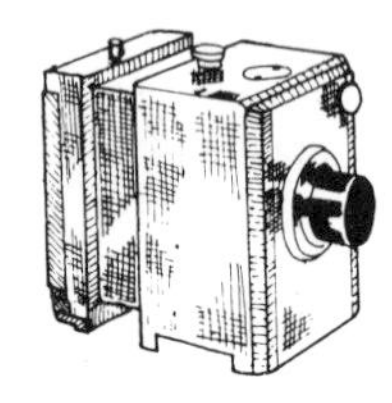

Rare Negretti and Zambra small wet-plate camera, circa 1860, 3¼in. square. $1,125 £500

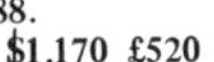

Stirn's small waistcoat detective camera, in brass, second model, circa 1888. $1,170 £520

A gigantic, art studio, 11½in. plate camera in mahogany, with a brass-cased J. H. Dallmeyer lens, on an ebonised tripod base. $1,225 £545

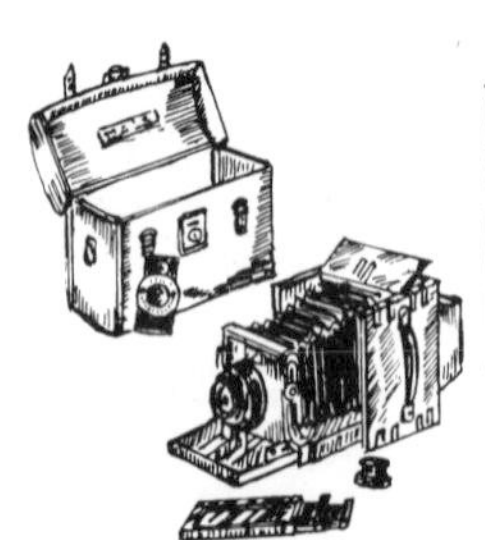

The 'Una' quarter-plate camera by James A. Sinclair. $1,250 £550

Large J. H. Dallmeyer tailboard studio camera, circa 1880. $1,240 £550

An Armanox high speed miniature press camera and case. $1,280 £570

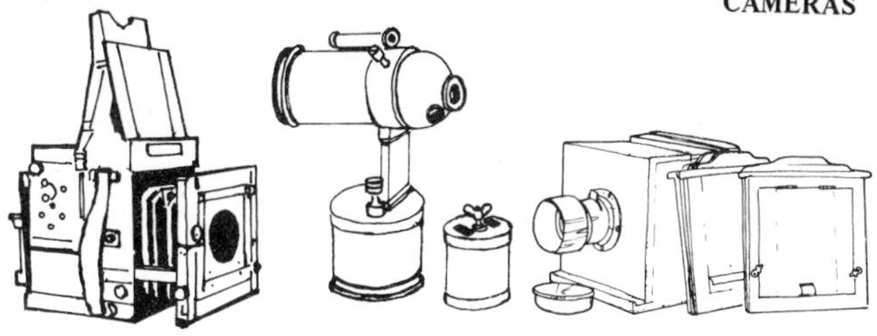

English Marion tropical reflex camera, circa 1920. $1,305 £580

Ferrotype 'mug' camera with spare developing tank, German, circa 1895, 11½in. high. $1,350 £600

A good wet plate camera, circa 1860. $1,410 £625

English wet-plate camera, 4 x 5in., circa 1860, with single element lens. $1,520 £650

'The Telephot' ferrotype button camera, circa 1900, 1ft.2½in. wide. $1,485 £660

J. Lizars 'Challenge' stereoscopic camera, circa 1900. $1,575 £700

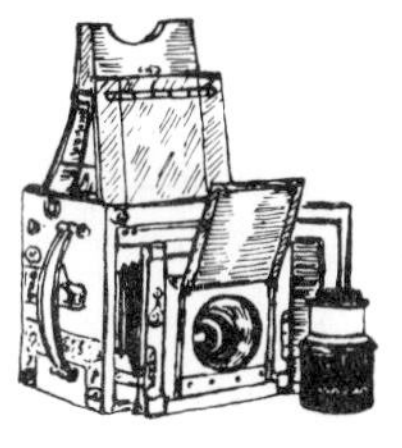

Marion type tropical reflex camera, circa 1920, in leather case. $1,855 £825

Early sliding box wet type or Daguerreo type camera, circa 1855. $1,870 £830

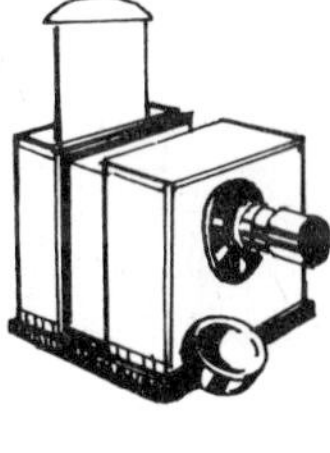

Wet plate camera, circa 1880, sold with screen and tripod. $1,890 £840

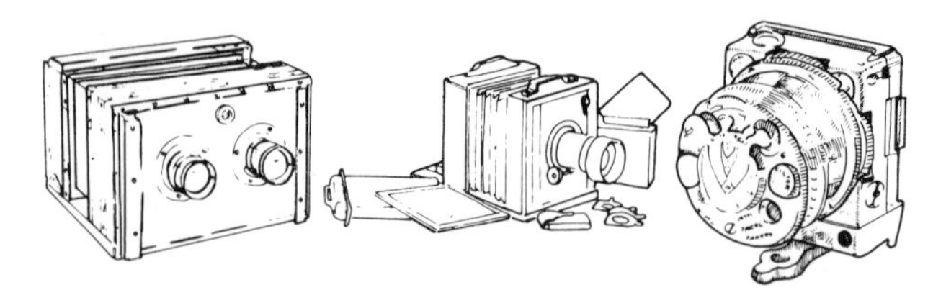

Brass bound Ross stereoscopic camera, circa 1870. $2,025 £900

Square bellows tailboard, 5 x 5in. wet plate Daguerreotype, about 1850. $2,025 £900

Compass camera by Jaeger Le Coultre. $2,080 £925

A sliding box Daguerreo type camera, circa 1850. $2,250 £1,000

Reddings Luzo mahogany roll-film camera in ever ready case. $2,475 £1,100

Rare black Leica IIIF 'red dial' camera, 36mm. long, circa 1956. $2,675 £1,100

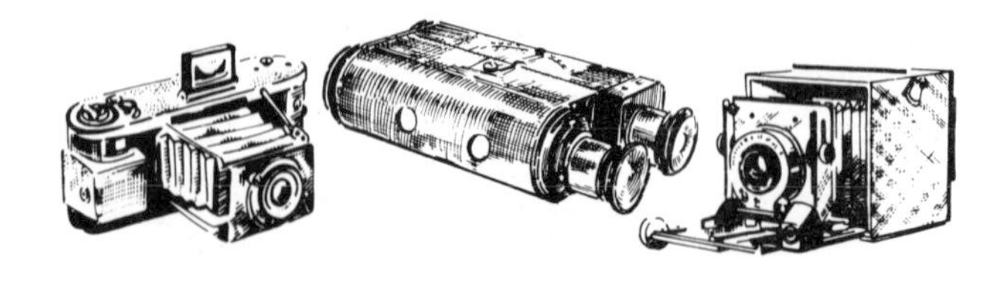

Palmos Jena focal-plane roll-film camera, 1902. $2,640 £1,100

W. Watson & Sons stereoscopic 'detective' binocular camera, circa 1900, 8in. high. $2,700 £1,200

Sinclair Una Traveller hand-and-stand camera. $2,735 £1,200

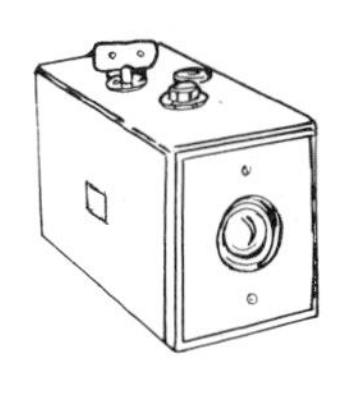

Kodak No. 1 roll-film camera in black leather case, 1888.
$3,600 £1,600

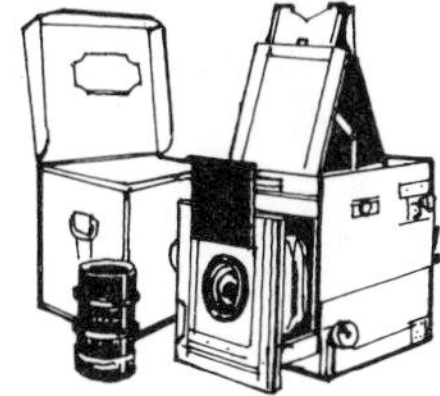

Early 20th century London Stereoscopic Co. improved artist's tropical reflex camera. $4,050 £1,800

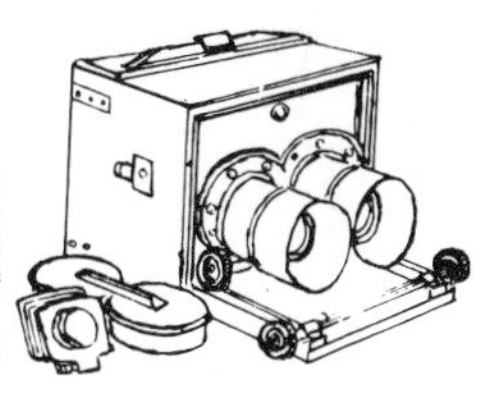

James How wet plate stereoscopic camera, English, circa 1880. $5,175 £2,300

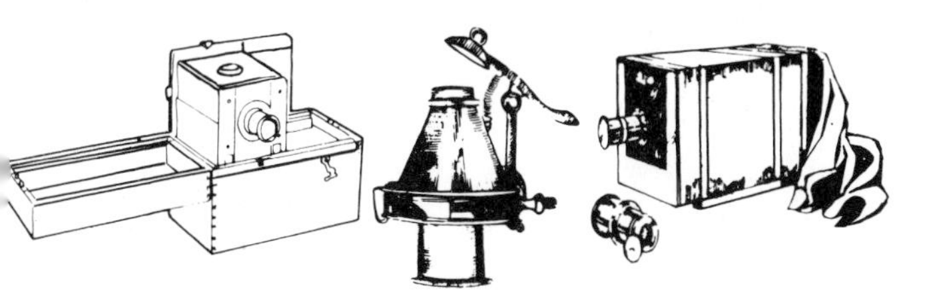

Powell's patent stereoscopic camera, 1858. $5,400 £2,400

Brin's patent spy-glass camera. $5,500 £2,500

An early Daguerreo type camera. $7,315 £3,250

Rare Johnson pantoscopic camera, circa 1860.
$9,000 £4,000

Rare Skaif's Pistolgraph miniature all-brass wet plate camera, around 1863.
$22,500 £10,000

A very rare, mid 19th century, Thomas Sutton wet plate camera.
$31,500 £14,000

19th century marine chronometer by John Bruce & Sons, 4¾in. diam. $900 £400

Marine chronometer, English, circa 1900, by Thomas Mercer Ltd., St. Albans, in mahogany case, 21cm. high. $1,015 £450

A 19th century ship's chronometer in a brass case by Thos. Hewitt, London, numbered 100, in a mahogany box marked Parkinson & Frodsham. $1,240 £550

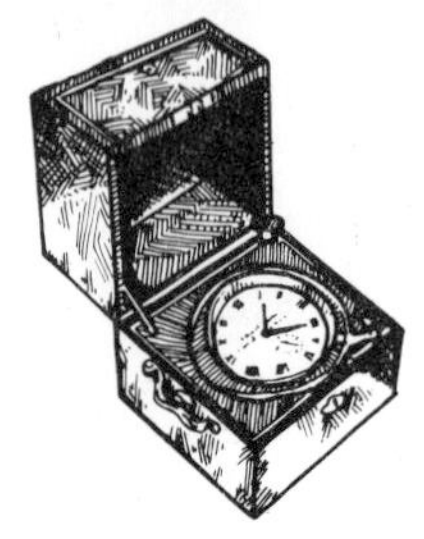

19th century marine chronometer by Wm. Bond & Son of Boston, 4½in. diam. $1,410 £625

Early 19th century two-day marine chronometer by Parkinson & Frodsham, London, 4¾in. diam. $1,430 £635

French chronometer, circa 1830, signed by Henri Robert, Paris, timepiece in leather case, with cylindrical movement, 6.7cm. long. $1,495 £665

An early 19th century, thirty-hour chronometer by Dent of London. $1,510 £670

A ship's chronometer by John Poole, London, numbered 3670, set in lacquered brass gimbals, in its brass mounted wooden case. $1,575 £700

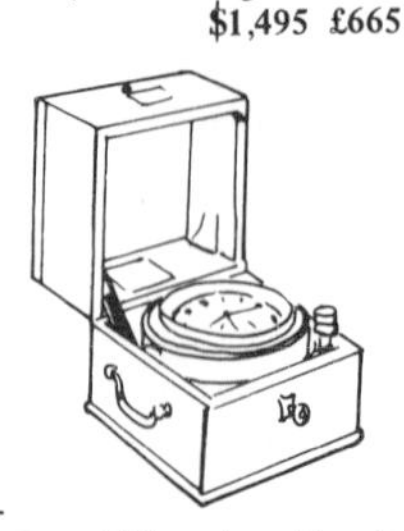

Late 19th century French Leroy two-day marine chronometer, 7in. high. $1,640 £730

Marine chronometer by James McCabe & Co., numbered 323, circa 1840. $1,690 £750

Ebonised mantel chronometer by T. Coombe, Brighton. $1,745 £775

French marine chronometer, circa 1850, by Vissiere Au Hayre, the silvered face having separate dials for hours, minutes and seconds, 21cm. high. $1,745 £775

Fine cased chronometer by Charles Frodsham. $1,910 £850

Two-day marine chronometer by Dent, London, dial, 10.5cm. diam. $2,230 £990

Fine cased marine chronometer by Arnold & Dent. $2,700 £1,200

A small eight-day marine chronometer signed on the full plate movement 'Barrauds'. $5,060 £2,250

Two-day marine chronometer by Breguet & Cie, dial 8cm. diam. $7,875 £3,500

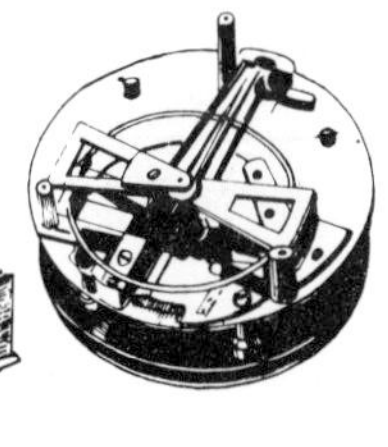

A fine marine chronometer, circa 1770. $9,560 £4,250

COMPASSES

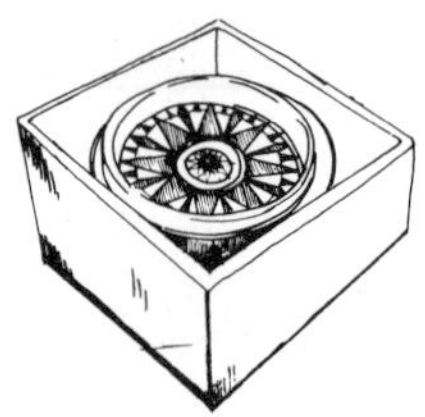

Ship's compass by Berry & Mackay, Aberdeen, in a wood box. $100 £45

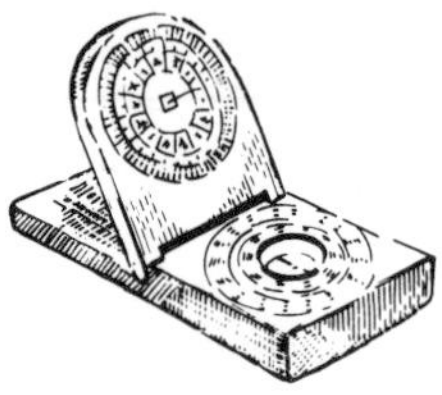

Chinese wood compass dial, 14.5cm. long. $115 £50

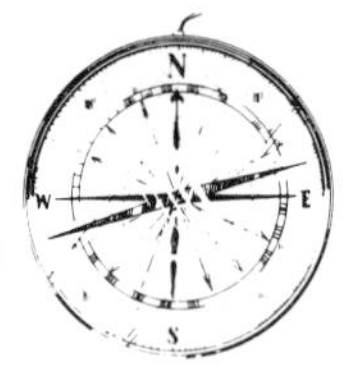

19th century mahogany cased compass. $115 £50

Surveyor's sighting crosshead and compass by W. S. Jones, London, circa 1840, with box. $295 £130

Table compass in the form of a gilt and silvered kettle drum. $340 £150

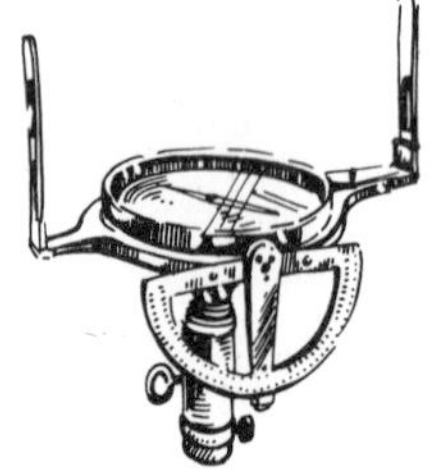

Mid 19th century English Davis & Son brass miner's compass, 10½in. wide. $360 £160

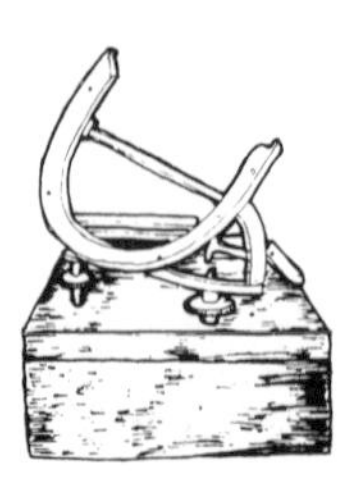

French compass sundial, circa 1830. $450 £200

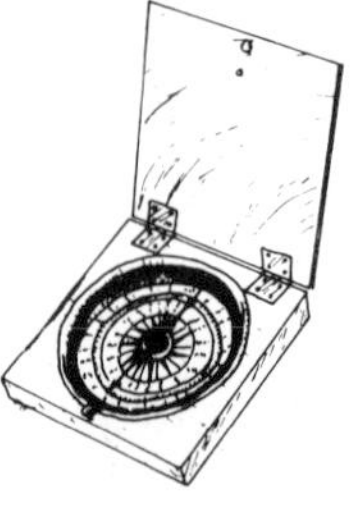

Early 18th century mahogany cased ship's compass by Thos. Wright, diam. of bezel 134mm. $690 £300

18th century French brass nautical compass and folding sundial. $1,690 £750

Victorian corkscrew with horn handle and brush, about 1860. $36 £16

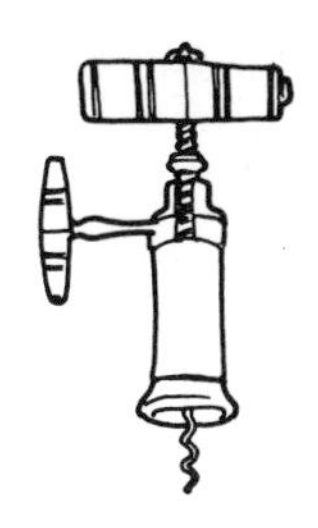

19th century Plum's patent ratchet corkscrew with ivory handle, brass barrel and steel screw. $90 £40

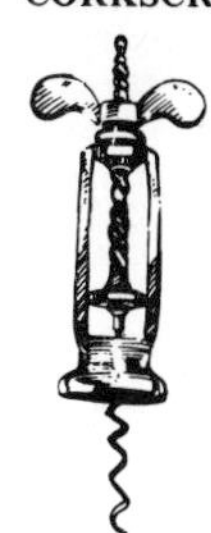

Unusual Victorian brass corkscrew. $125 £55

Victorian brass and ivory corkscrew. $205 £90

Unusual brass Victorian corkscrew. $360 £160

Late 18th century example of the first patented corkscrew in this country. $405 £180

Steel rack and pinion hand corking device, 6¼in. high. $620 £275

Steel corkscrew by Chas. Hull, Birmingham, 1864. $675 £300

Dutch silver corkscrew by Hendrik Smook, Amsterdam, 1753. $1,465 £650

DIALS

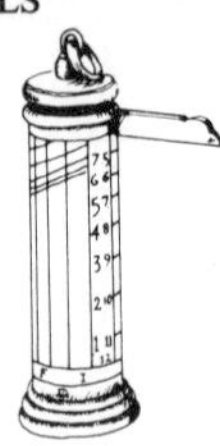

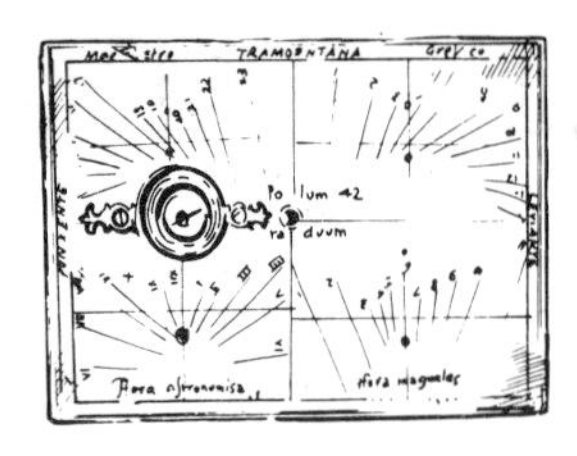

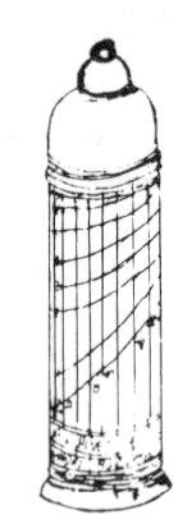

18th century brass pillar dial, inscribed Thomas Miller, Knowle, 9.5cm. high. $1,000 £420

Fine Italian gilt metal Pingnomon horizontal dial, 15.8cm. long. $4,210 £1,870

17th century ivory pillar dial. $3,265 £1,450

COMPENDIUMS

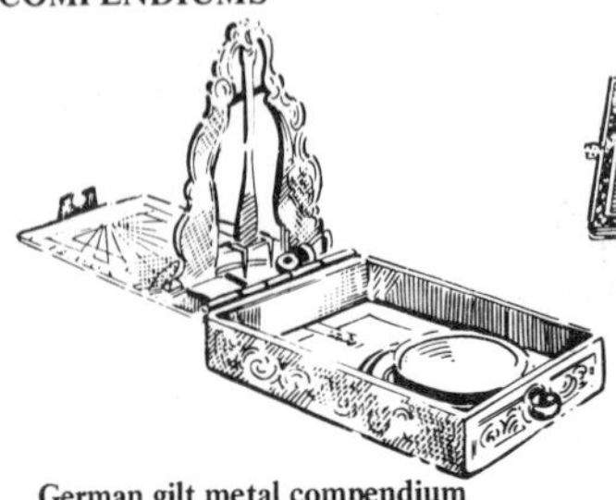

Folding pocket compendium by Christopher Schissler, Augsburg, in gilt metal, 1575.

$11,250 £5,000

German gilt metal compendium dial. $7,425 £3,300

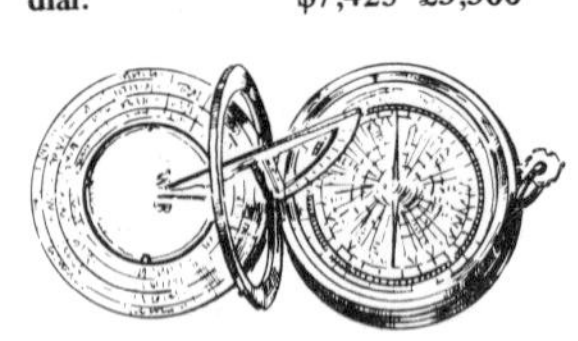

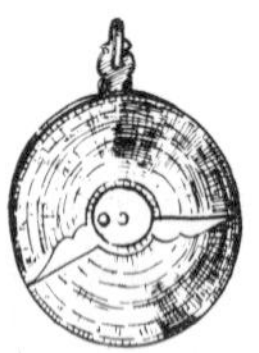

Early 17th century gilt metal compendium by Elias Allen, 63mm. diam. $18,560 £8,250

An important compendium by Christopher Schissler of Augsburg. $33,750 £15,000

DIPTYCH DIALS

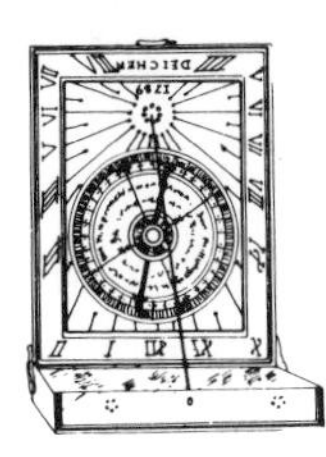

German fruitwood diptych dial, signed and dated Deicher, 1789, 10.7cm. high. $1,555 £650

17th century ivory diptych dial signed Fait et Inv. par Charles Bloud a Dieppe, 8cm. long. $1,800 £800

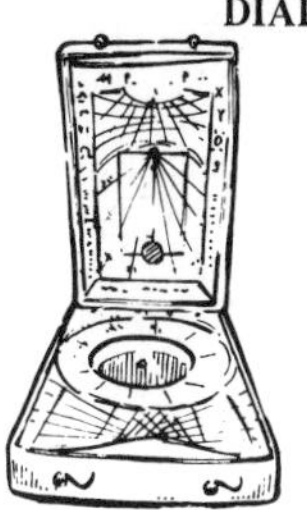

German ivory diptych dial by Hans Troschel. $3,710 £1,650

EQUINOCTIAL

German gilt metal equinoctial dial, 65mm. diam. $640 £285

18th century German cased engraved brass and steel equinoctial sundial, 6cm. square. $850 £380

Brass equinoctial dial by Pizzala, London, 13.5cm. diam. $990 £440

Brass universal equinoctial dial by Troughton & Simms, London, 13.5cm. diam. $1,125 £500

German silvered and gilt metal universal equinoctial dial by And. Vogl, 7.2cm. greater dimensions. $1,195 £500

English universal equinoctial dial, 4in. diam. $1,465 £650

DIALS
EQUINOCTIAL

Brass equinoctial dial by Fraser & Sons, London. $1,575 £700

17th century equinoctial ring dial and case by D. Hendrickson, 1650. $2,700 £1,200

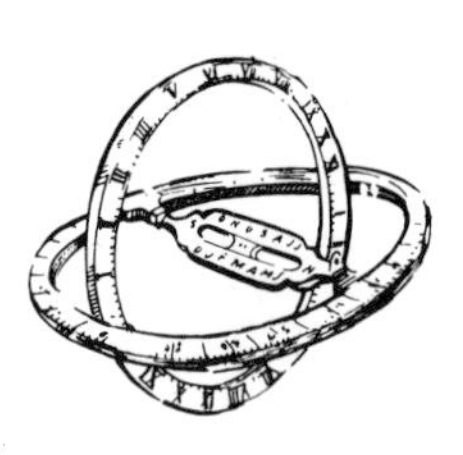

Fine large universal equinoctial ring dial, 9in. diam. $3,375 £1,500

EQUATORIAL

An 18th century circular universal equatorial dial by Adams. $1,125 £500

Early 18th century brass universal equatorial dial, 63mm. diam. $1,575 £700

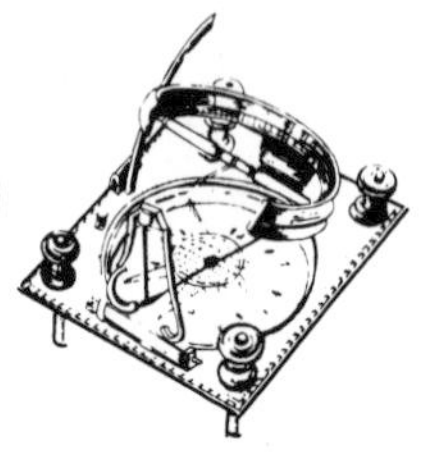

Russian brass equatorial dial, circa 1780. $1,690 £750

NOCTURNAL

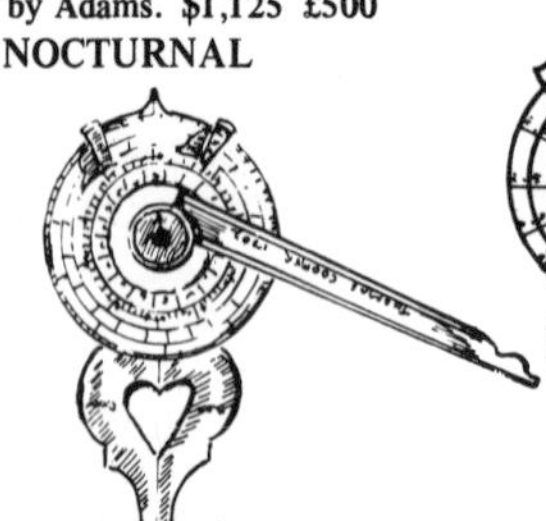

18th century boxwood nocturnal dial, inscribed 'Thomas Cooper 1701'. $1,910 £850

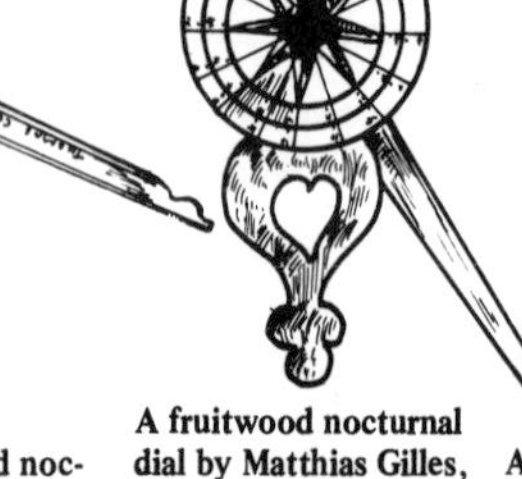

A fruitwood nocturnal dial by Matthias Gilles, dated 1708, overall length 262mm. $2,250 £1,000

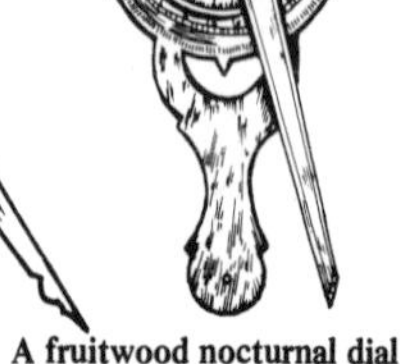

A fruitwood nocturnal dial signed William Broughton and dated 1679. $2,925 £1,300

English brass universal equinoctial ring dial, unsigned, 8.8cm. diam.
$860 £360

Universal ring dial by Troughton & Simms, London.
$1,575 £700

Mid 18th century brass ring dial by Claude Langlois, 100mm. diam.
$2,455 £1,050

TABLET

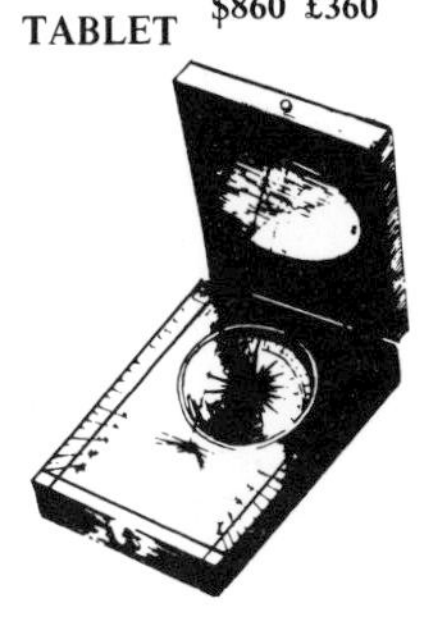

19th century English mahogany tablet dial, 3in.
$170 £75

Ivory magnetic analemmatic dial signed C. Bloud a Dieppe, 3¾in. wide. $2,080 £925

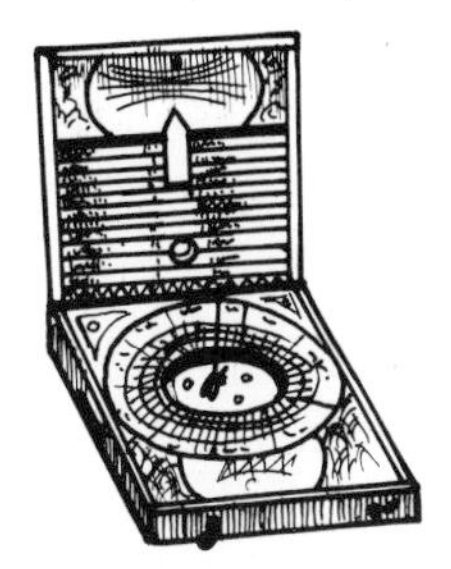

Ivory tablet dial signed Leonhart Miller, 1636, 4in. long.
$2,250 £1,000

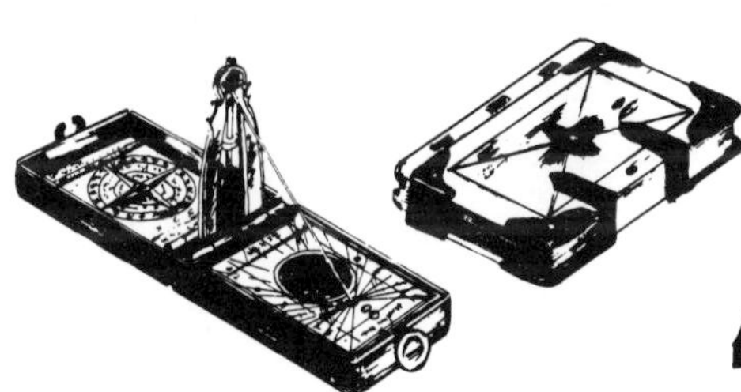

A gilt tablet dial by Tobias Volchmer, circa 1612.
$2,530 £1,125

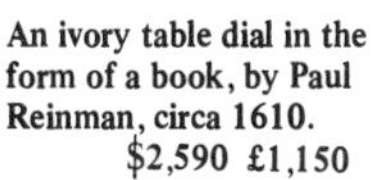

An ivory table dial in the form of a book, by Paul Reinman, circa 1610.
$2,590 £1,150

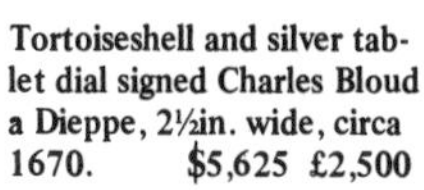

Tortoiseshell and silver tablet dial signed Charles Bloud a Dieppe, 2½in. wide, circa 1670. $5,625 £2,500

DIVIDERS

Large, wheelwright's dividers, circa 1820. $80 £35

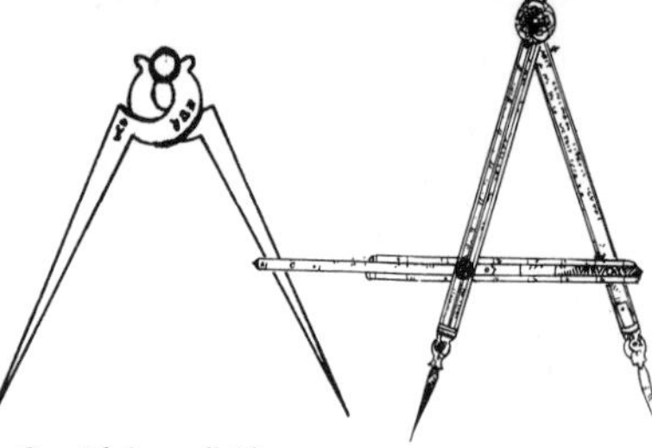

Spanish brass dividers, circa 1585, length of arms 250mm. $900 £400

A fine pair of gilt metal dividers by Christopher Schissler. $13,500 £6,000

DOMESTIC INSTRUMENTS

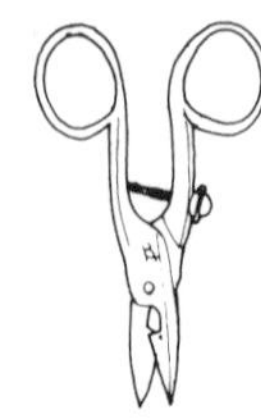

Steel buttonhole scissors, made in Sheffield by Walker & Hall, about 1920. $6 £3

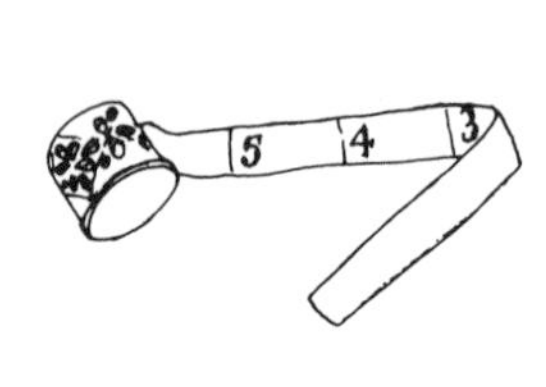

Miniature bone-cased tape measure. $10 £5

American, Dover pattern whisk, in iron, circa 1904. $15 £7

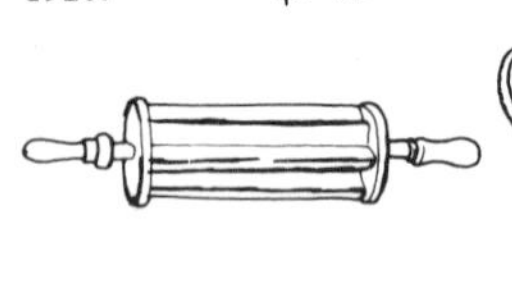

Butter worker used on a sloping tray. $18 £8

Unusual late Victorian solid steel scissors with swan decoration. $18 £8

19th century wooden needle case and thread spool. $20 £9

Small Victorian chopper, with wooden handle. $20 £9

Brass pie trimmer and wheel, circa 1830. $22 £10

Victorian brass and copper horse mane singer. $36 £16

Steel sugar cutters, circa 1790. $45 £20

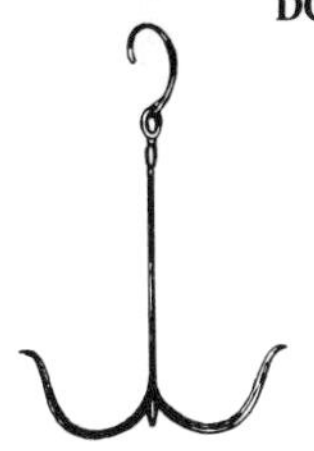

Queen Anne period steel game hook. $55 £25

A small 19th century butcher's chopper in steel with turned oak handle, 13in. long. $55 £25

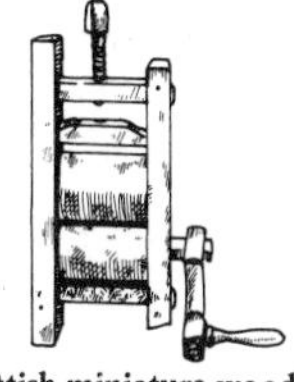

Scottish miniature wooden mangle for clerical bands. $55 £25

Exceptionally large old butcher's cleaver, 26in. long. $70 £30

Georgian silver framed spectacles by E. T., London, 1823. $70 £30

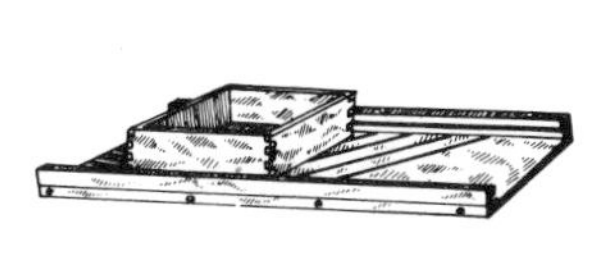

Large, Victorian kraut cutter of wood and brass. $80 £35

Swedish steel vegetable chopper. $90 £40

A brass and steel desk knife sharpener, with steel roller supports on pillars, circa 1825, 5½in. long, 4in. deep and 3½in. high. $90 £40

A collection of 19th century hand-wrought flesh forks with wall bracket. $90 £40

Brass and cow horn pocket fleam, circa 1820. $95 £40

Queen Anne wrought iron gridiron with four feet to stand on embers, circa 1700, 9in. square. $100 £45

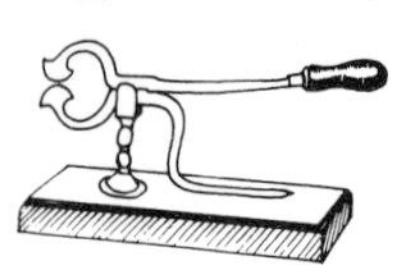

Pair of 18th century brass sugar cutters on a mahogany stand. $100 £45

Brass 'Hemming Bird' surmounted by a pincushion. $100 £45

A set of three fire implements with Adam style handles. $100 £45

Late 19th century adjustable hat stretcher of walnut and cast iron, 13in. high. $108 £48

17th century steel steak fork, ram's horn top, circa 1660. $125 £55

Antique brass door knocker, 8in. high, with circular brass plate. $108 £48

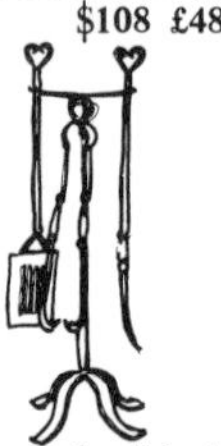

A fine set of polished steel fire irons on a stand, circa 1840. $160 £70

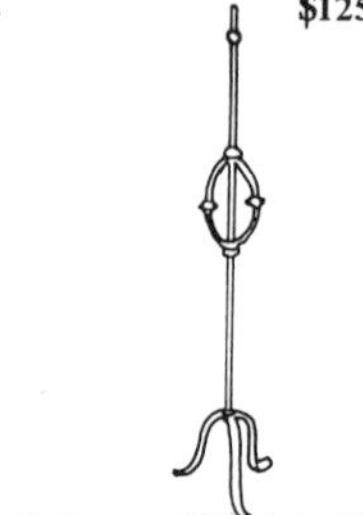

A rare polished steel Lark-spit, 30½in. high, circa 1750. $160 £70

Pratts ethyl gasolene lighter fuel dispenser. $180 £75

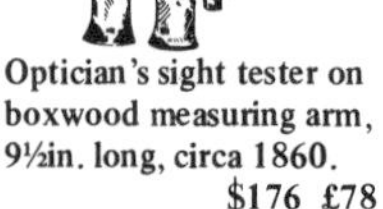

Optician's sight tester on boxwood measuring arm, 9½in. long, circa 1860. $176 £78

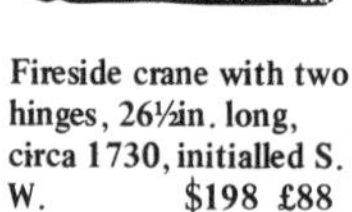

Fireside crane with two hinges, 26½in. long, circa 1730, initialled S. W. $198 £88

18th century plate warmer. $215 £95

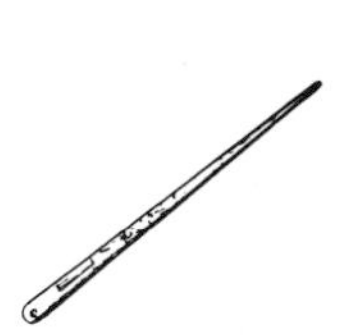

A bodkin by G. W., London, circa 1660, 10½cm. long. $235 £105

Butcher's cast brass bell-shaped scale weights, circa 1850, graduated sizes from 7lb. –½oz., nine in all. $235 £105

Early 18th century iron hanging rushlight and candleholder, 4ft. long, extending to 6ft. $250 £110

Queen Anne rushlight holder, 36in. high. $350 £155

Pair of 19th century walnut nutcrackers. $370 £165

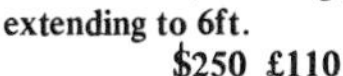

Pair of 17th century brass tongs, 13.5cm. long. $455 £190

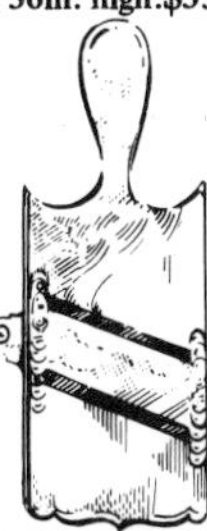

Early 19th century Dutch cucumber slicer, mounted on boxwood. $450 £200

Early 19th century brass winding bellows. $430 £200

Adjustable, Queen Anne period, wrought iron rushlight and candle stand, circa 1700. $450 £200

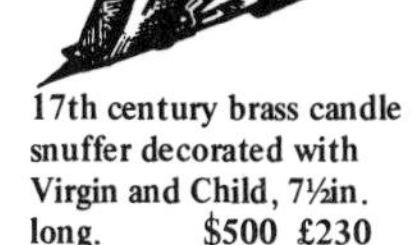

17th century brass candle snuffer decorated with Virgin and Child, 7½in. long. $500 £230

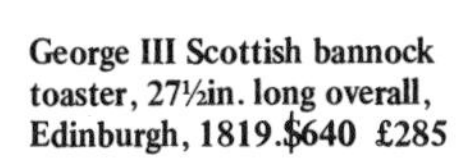

George III Scottish bannock toaster, 27½in. long overall, Edinburgh, 1819. $640 £285

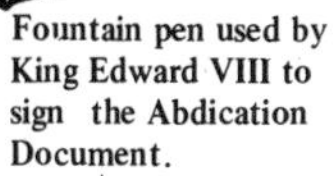

Fountain pen used by King Edward VIII to sign the Abdication Document. $4,500 £2,000

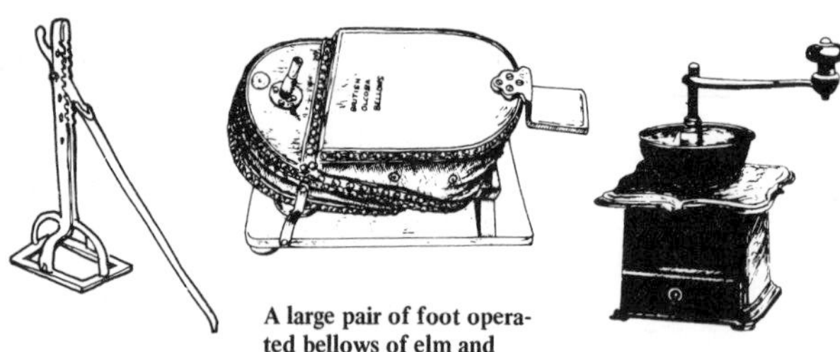

Adjustable iron cart jack, 19th century. $35 £15

A large pair of foot operated bellows of elm and leather construction, on baseboard, 1ft.9in. long, 1ft.2in. high. $55 £25

Mid Victorian coffee grinder. $55 £25

Cast iron and brass Victorian mincer by Burgess & Key, 13 x 6in. $115 £50

Copper fire extinguisher with five oval brass plaques, 1930, 29in. high. $125 £55

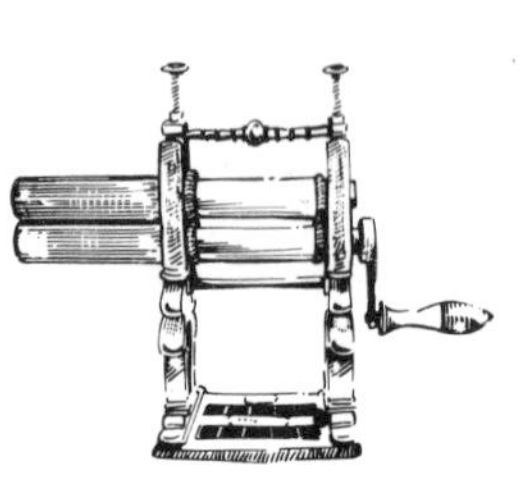

Late 19th century patent crimping machine. $140 £60

Stained oak spinning wheel, circa 1880, 3ft.4½in. high. $180 £80

An interesting farmhouse cheese press, with wrought iron frame, 20in. high, circa 1800. $180 £80

Late 19th century mechanical bellows in black painted cast iron and brass machine, 23½in. high. $170 £90

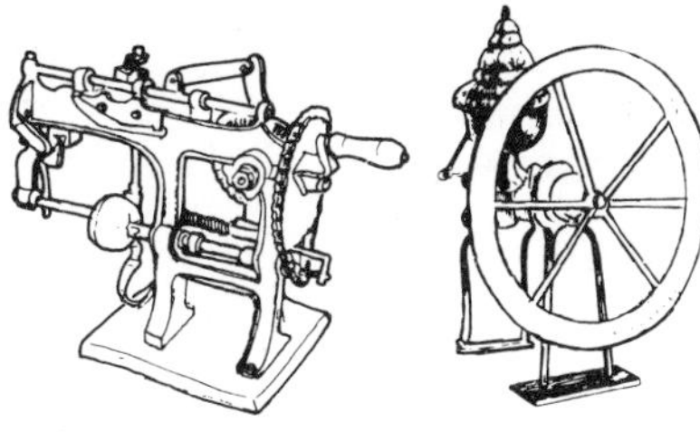

Apple corer made around 1890-1910, works on the same principle as a sewing machine. $280 £125

A fine Victorian brass and steel coffee grinder, 26in. high. $295 £130

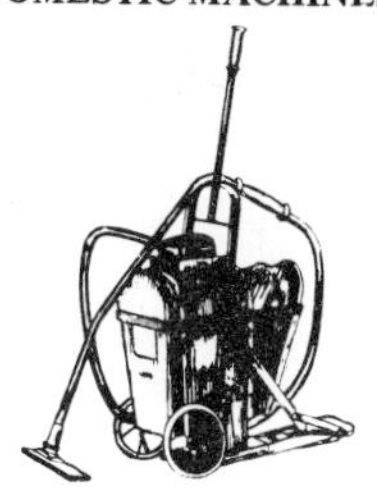

An early English vacuum-cleaner, with iron-spoked wheels and rubber tyres, 3ft.10in. high, circa 1900. $280 £125

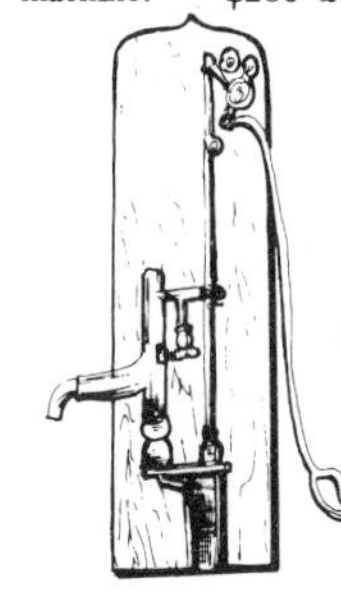

Highly polished brass, steel and lead farmhouse kitchen water pump. $295 £130

English sweet-making machine in cast iron, with brass scoop and copper pan, circa 1880's. $420 £180

Coffee merchant's grinder with original paintwork, 27in. high. $450 £200

18th century oak coffee mill, 7¼in. high, with brass mounted grinder handle. $620 £275

18th century hand wrought steel cheese press, complete with large brass bowl, 56in. high, circa 1720. $790 £350

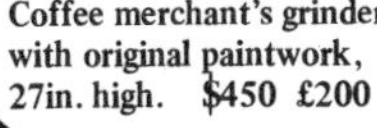

Small hand pump which belonged to the Birmingham Fire Office. $1,070 £475

Early 19th century French padlock with oak leaf decoration on key escutcheon cover. $36 £16

17th century bar padlock and key. $100 £45

George III brass door lock, 7in. wide, with keeper and key. $135 £60

17th century engraved door lock and key. $145 £65

French steel-cased door lock with heavy ormolu trim, brass handles and escutcheon plate, circa 1790. $225 £100

Louis XVI ormolu door lock. $340 £150

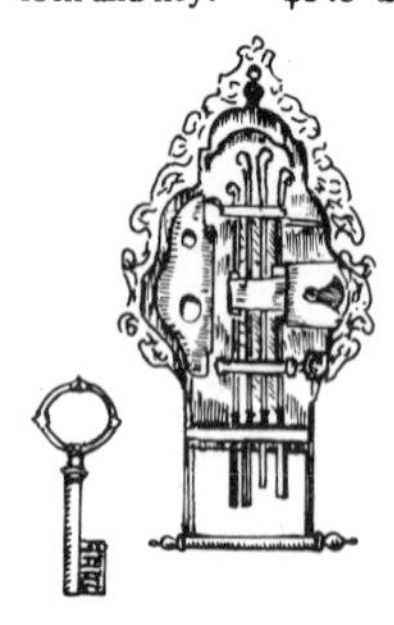

Late 17th century German door lock with brass finials. $450 £200

Late 17th century German door lock in iron, 13in. long, with original key. $1,240 £550

16th century steel door lock with key, from the Bohemia's Castle of Dux, where Casanova's body was finally laid to rest, 19½in. wide. $1,410 £625

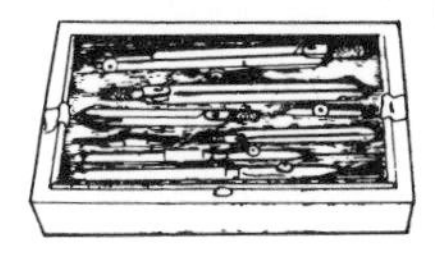

A set of architect's drawing instruments by Thornton Ltd., in mahogany case. $80 £35

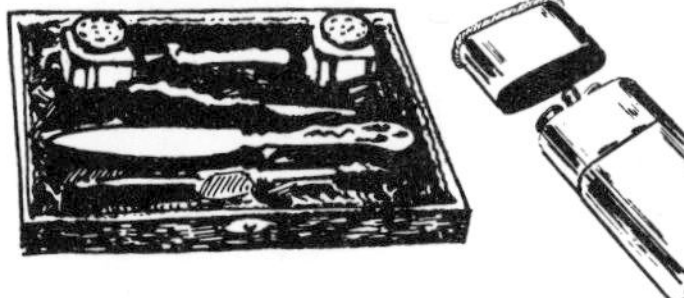

Silver mounted writing set, comprising letter opener, pen and knife, seal and two ink bottles. $90 £40

George IV travelling writing set, London, 1826, maker A. D., 3¼in. long. $215 £95

19th century desk set containing pen, paper knife and seal, 8½ x 3½in. $235 £105

Early 19th century set of drawing instruments, 5in. long. $250 £110

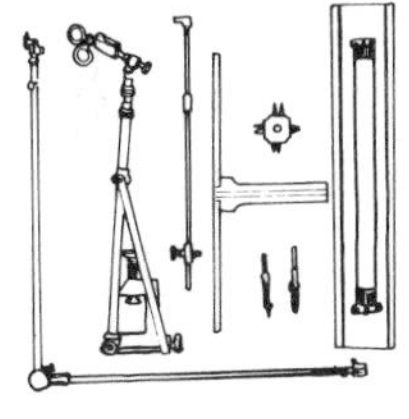

Late 19th century set of Stanley drawing instruments, in fitted wooden case. $295 £130

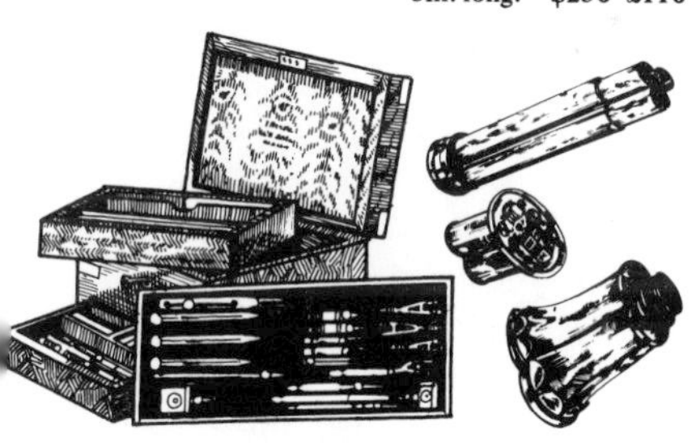

A good set of early 20th century drawing instruments by J. Helden & Co. $350 £155

A fine silver mace-shaped penner, circa 1690, with slots for three quills, maker W. B. $900 £400

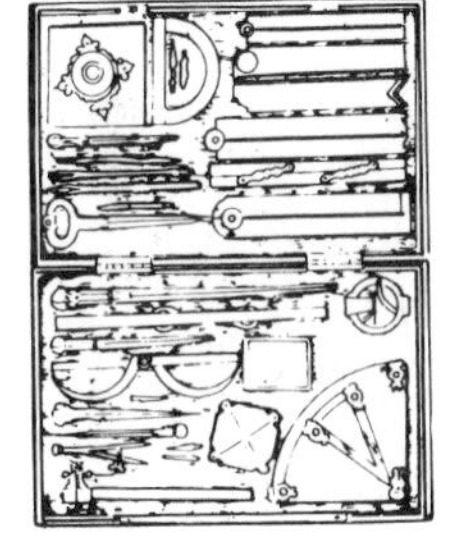

Rare set of mathematical instruments by Dominicus Lusuerg, Rome, 1701. $36,000 £16,000

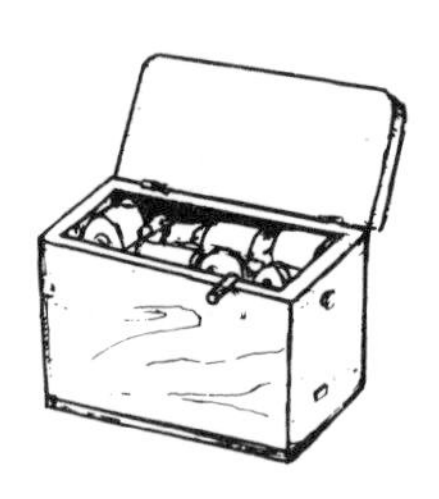

Victorian magneto electric machine in an oak case. $55 £25

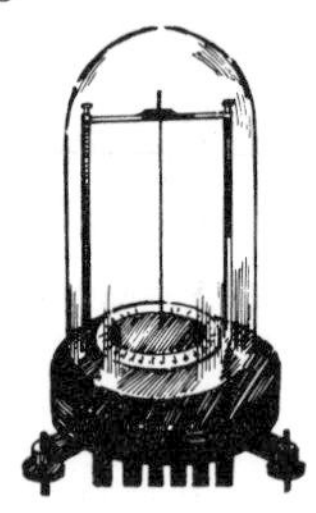

Late 19th century galvanometer. $160 £70

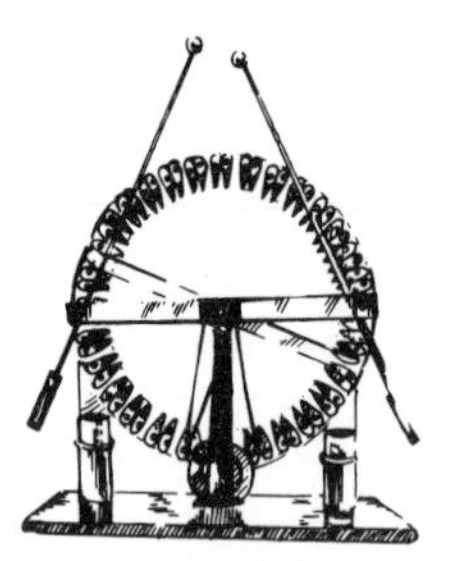

Early 20th century English Griffin & Tatlock Wimshurst machine 1ft.3½in. high. $200 £90

Wimshurst machine for producing electricity, 14ins. high. $295 £130

Set of electrical apparatus, including galvanometer, induction coil, wet cell battery, etc., in wooden case, English, 1880-1900. $505 £220

A large early 20th century electric influence machine, 2ft.2in. wide, contained in a wooden case. $450 £200

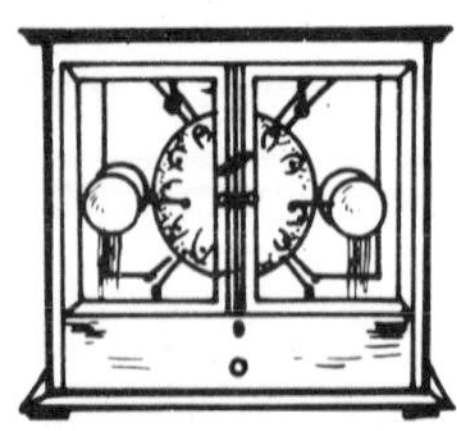

A Wimshurst electric influence machine with three pairs of amber discs, labelled Griffen, circa 1900, 2ft.2in. wide. $700 £310

Mahogany brass and glass Wimshurst experimental electric machine, circa 1880. $775 £345

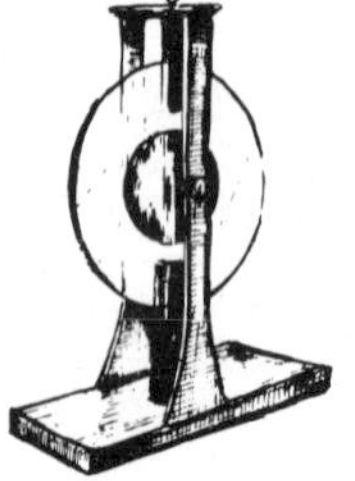

A rare, early electrostatic friction machine by J. Cuthbertson, 1799. $1,070 £475

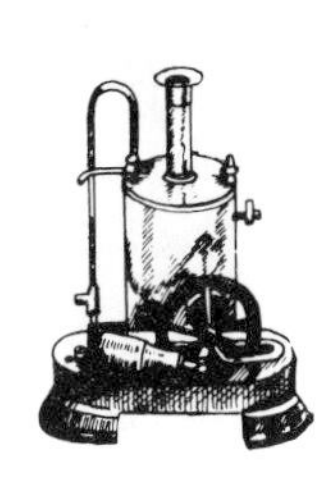

A working model of a piston with flywheel, 8in. high. $80 £35

Small vertical single cylinder stationary steam engine, circa 1920, 9¼in. high. $115 £50

Good Doll & Co. stationary steam plant with spirit burner, 1922. $115 £50

A model of a horizontal stationary steam engine with large brass boiler, mounted on a stand, 8½in. by 4in. $170 £75

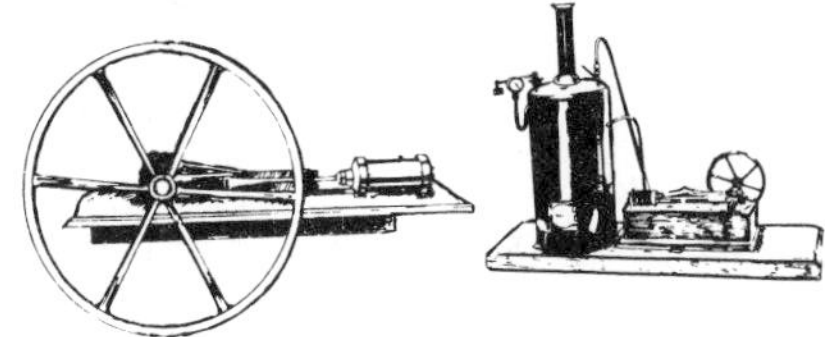

Late 19th century brass horizontal model mill engine. $190 £85

19th century model of a horizontal steam engine. $190 £85

Victorian model of a stationary steam engine. $190 £85

Stuart, vertical, double-action steam engine, 16½in. high. $200 £90

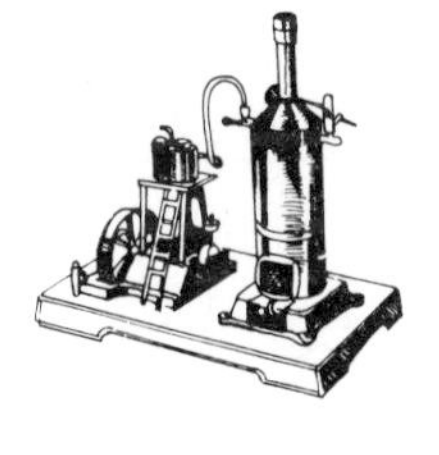

A model of a vertical stationary steam engine with large brass vertical boiler, 13in. high. $215 £95

A Victorian coal fired, model steam engine, 13½in. high. $295 £130

Early model gas engine, 10in. long, circa 1920's. $315 £140

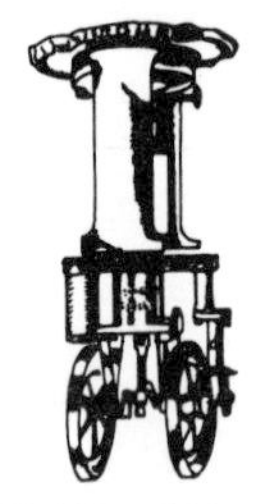

'Heinrici' model hot air engine, 18in. high. $395 £175

Marklin industrial steam plant, circa 1915. $540 £240

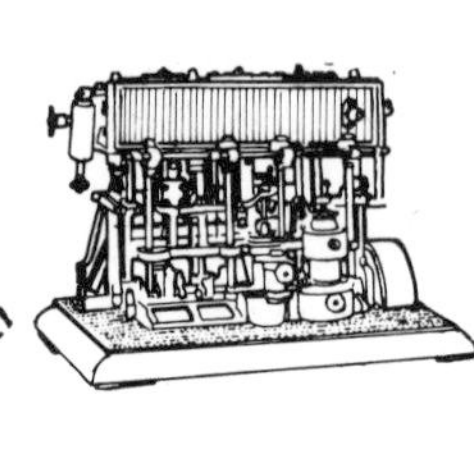

Fine model of a triple expansion marine steam engine, finished in grey and mounted on a chequered baseboard, 10¾in. long. $790 £350

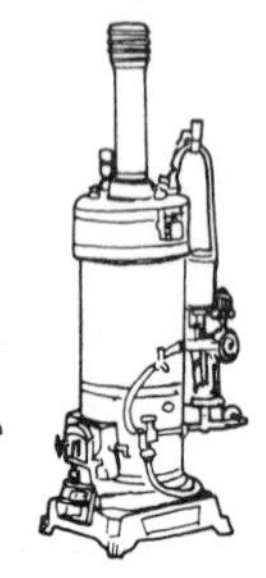

Fine Bing vertical boiler and steam engine, 1ft.4¼in. high, circa 1930's. $900 £400

Fine apprentice made model of a ship steam engine, 10 x 15in. $900 £400

Good 19th century vertical twin-cylinder stationary steam engine in brass and steel, 1ft.3¼in. high, the valve inscribed G. H. Joslin, 1880. $955 £425

Late 19th century model twin-cylinder horizontal stationary mill engine, 1ft. 3½in. long. $1,690 £750

Rococo etui, chased and embossed and with gilt fittings. S135 £60

English etui, circa 1760, 11cm. long. $295 £130

Mid 18th century gilt metal and hardstone etui case, 11cm. long. $395 £175

Bilston enamel rainbow etui case, 4½in. high, circa 1765-70. $620 £275

George II etui, circa 1745, in green shagreen case. £675 £300

Gold mounted shagreen and enamel etui, late 18th century, Swiss 4in. high. $790 £350

Gilt metal etui on a chatelaine with dark and light brown agate pendants. $945 £420

Continental gilt metal etui with some fittings. $970 £430

French etui with six fittings, including a knife, scissors, pencil, ivory slide, spoon and bodkin, circa 1740. $2,250 £1,000

Brass and mahogany reel, 4½in. diam., circa 1840. $45 £25

Brass and wooden reel by Wilkes Sprey Brand, 3in. diam. $45 £25

Brass and mahogany salmon reel, circa 1820, 4in. diam. $80 £35

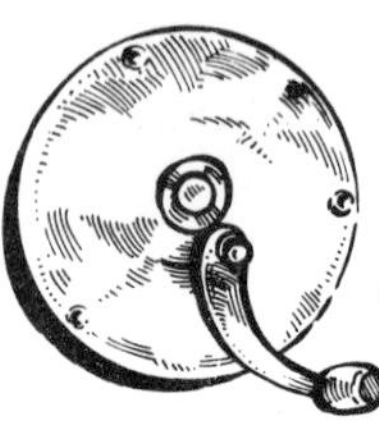

19th century all brass trout reel, circa 1810, 2¼in. diam. $90 £40

Fine 19th century solid brass trout reel, 2½in. diam. $90 £40

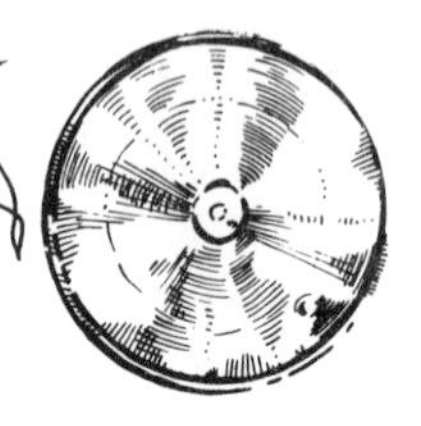

Fine cow horn and brass fishing reel, circa 1830, 4½in. diam. $100 £45

Fine, fisherman's salmon gaff, 16½in. long circa 1840. $100 £45

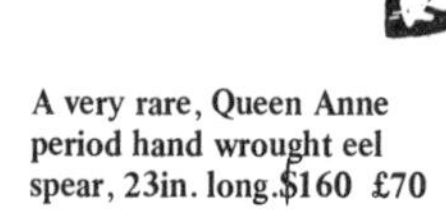

A very rare, Queen Anne period hand wrought eel spear, 23in. long. $160 £70

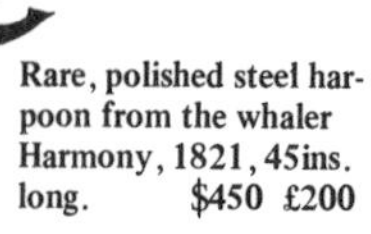

Rare, polished steel harpoon from the whaler Harmony, 1821, 45ins. long. $450 £200

19th century globe on iron stand, 20in. high. $270 £120

Newton, Son & Berry celestial globe, circa 1840, 10in. diam. $295 £130

A Donaldson's celestial globe, dated 1830, engraved by W. & A. K. Johnston, 11½in. high. $340 £150

Celestial globe by Newton, 1850, 12in. diam. $700 £310

A fine, early 19th century globe with a mahogany stand, 2ft.7ins. high. $900 £400

An important celestial globe by Cary, dated 1790, 52ins high. $1,910 £850

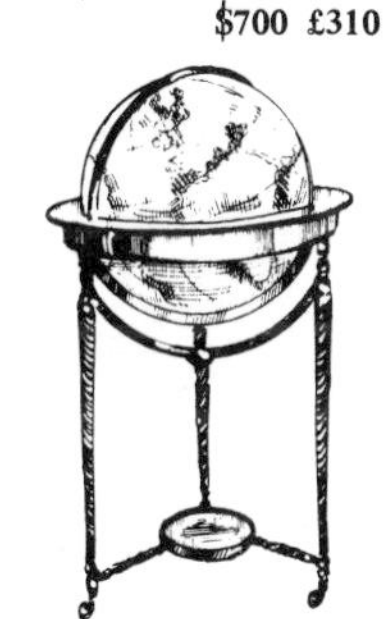

George III celestial globe by Dudley Adams, 18ins. diameter. $2,925 £1,300

George III Cary's celestial globe, 1799, 2ft.4in. diam. $3,940 £1,750

Fine example of John Russell's selenographia, 1797, 517mm. high. $29,250 £13,000

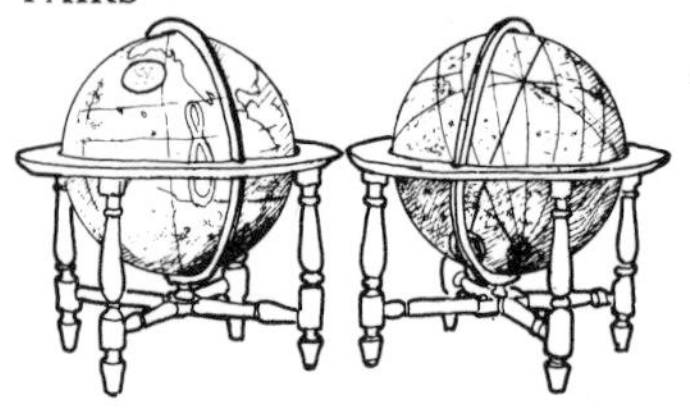

Pair of Cary's terrestrial and celestial table globes, 1825.
$2,590 £1,150

Pair of terrestrial and celestial globes by W. & J. Cary, London, on splayed tripod bases incorporating compasses in the stretchers. $3,375 £1,500

A fine pair of early 19th century celestial and terrestrial globes on tripod bases.
$3,375 £1,500

A pair of Cruchley's late Cary's terrestrial and celestial globes, 34in. high, 1840's. $3,710 £1,650

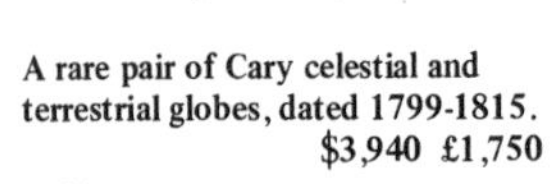

A rare pair of Cary celestial and terrestrial globes, dated 1799-1815.
$3,940 £1,750

Pair of Cary celestial and terrestrial dated globes, each with a compass in the stand.
$4,950 £2,200

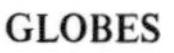

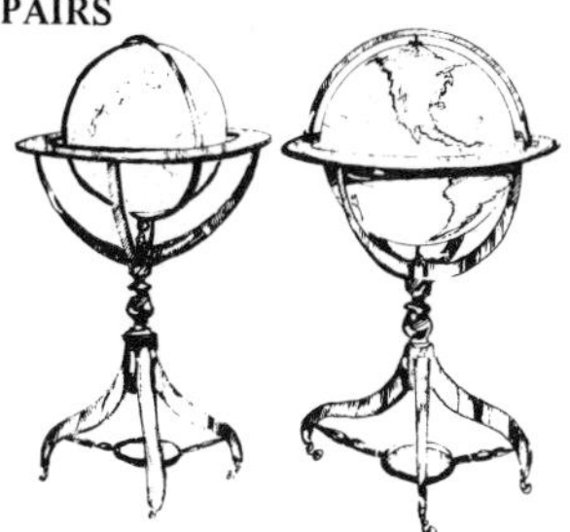

Pair of 19th century celestial and terrestrial globes.
$5,175 £2,300

A pair of 19th century terrestrial and celestial globes on mahogany stands.
$5,850 £2,600

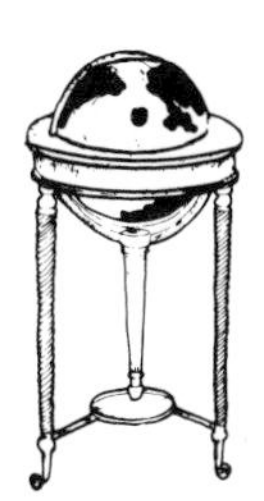

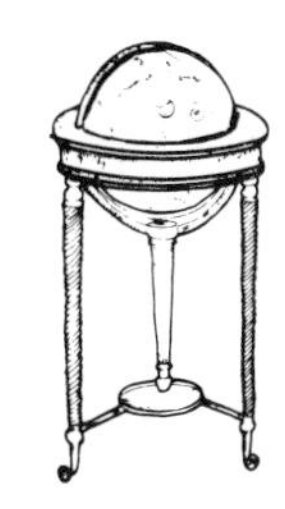

A pair of Regency library globes on mahogany stands, by William Harris & Co., 4ft. high, 2ft. diam.
$7,425 £3,300

A pair of George IV Cary's terrestrial and celestial globes, 48in. high.
$9,000 £4,000

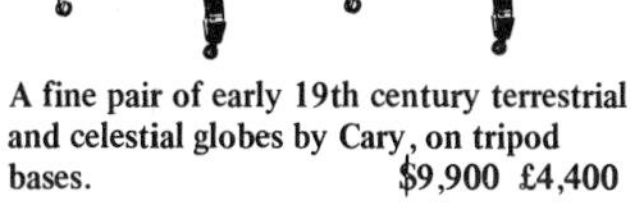

A fine pair of early 19th century terrestrial and celestial globes by Cary, on tripod bases. $9,900 £4,400

An important pair of Venetian celestial and terrestrial globes by Vincenzo Coronelli, 42in. diam., circa 1690.
$22,500 £10,000

GLOBES
POCKET

An unusual late 18th century small terrestrial globe in a shagreen case, the interior with a celestial globe, 6cm. diameter. $450 £200

Newton's New and Improved Terrestrial Pocket Globe, 1817. $570 £250

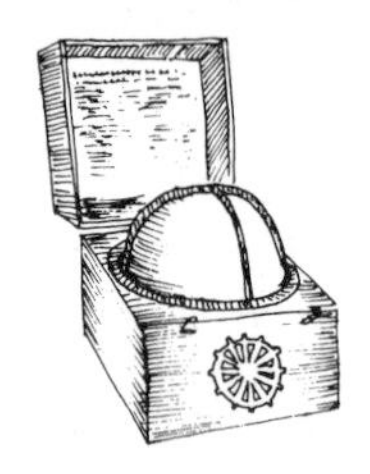

A ship navigator's star globe by H. Hughes & Son, London. $890 £385

Pocket terrestrial globe in a green shagreen case, 3in. diameter. $1,125 £500

Black sharkskin covered pocket globe by Newton showing a celestial map. $1,240 £550

A pair of late 18th century celestial and terrestrial pocket globes by J. & W. Cary, April 1791, 3in. diameter. $1,800 £800

TERRESTRIAL

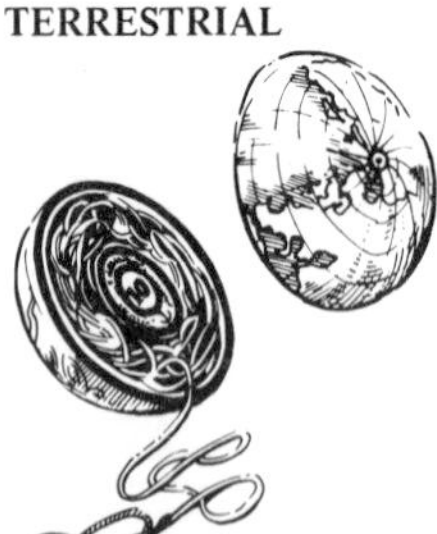

Model of a terrestrial globe in the form of a string box, circa 1890, 4¼in. high. $190 £85

Terrestrial globe on a mahogany turned stand. 15in. high, 1877. $405 £180

19th century globe on metal and brass stand. $450 £200

Manning & Wells terrestrial globe, 20in. diam., circa 1854. $450 £200

Regency terrestrial globe in a beechwood frame, 14ins. diameter. $550 £245

Sheraton mahogany terrestrial globe, circa 1825, 18in. high. $620 £275

Late George III mahogany terrestrial globe, by H. & T.M. Bardin, 2ft. diam., circa 1800. $2,810 £1,250

19th century terrestrial globe, 24in. diameter. $2,880 £1,280

Mid 19th century Johnston 30in. terrestrial globe, 47 x 40in. $3,340 £1,485

Regency mahogany terrestrial globe, 2ft. 2in. diam., by Cary's London, 1815. $3,825 £1,700

Large terrestrial globe by Smith of London, in a mahogany stand. $4,330 £1,900

Regency terrestrial globe by Newton, London, 1ft.9in. diam., circa 1820. $4,330 £1,900

KALEIDOSCOPES

A London Stereoscope Co. jewel kaleidoscope, circa 1860. $350 £155

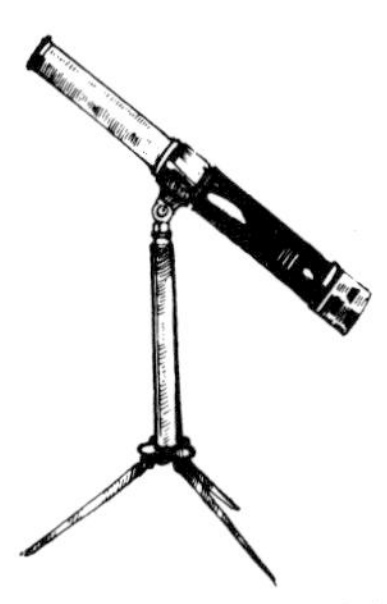

Early 19th century brass kaleidoscope by William Harris, 1ft. long. $595 £265

KEYS

Early 19th century iron key, 5in. long. $15 £7

Two large early iron keys. $28 £12

Gigantic George III iron door key. $55 £25

17th century key, the head with geometric motif, 4½in. long. $100 £45

A small 18th century cut steel key, 4in. long. $135 £60

17th century steel key with geometric motif, 5½in. long. $145 £65

An 18th century cut steel key, 5¼in. long. $235 £105

An 18th century cut steel key with a comb end, 5½in. long. $280 £125

Early 17th century French steel key, 5½in. long. $1,485 £660

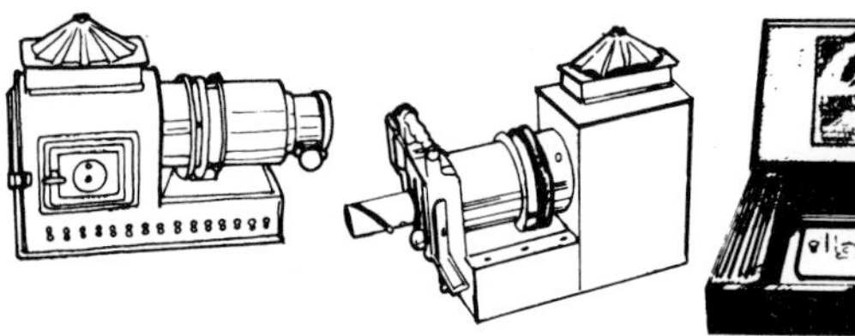

Baird lantern slide projector (enlarger) with electric light and wood tripod stand. $90 £40

Baird magic lantern, the slide with coloured plates. $90 £40

A German Lanterna Magica with a quantity of slides in case. $115 £50

Mahogany and brass lantern converted to electricity. $135 £60

Early magic lantern. $140 £62

Late 19th century English double projection lantern, 26in. high. $450 £200

Double oxy-hydrogen biunial limelight lantern in original case. $475 £210

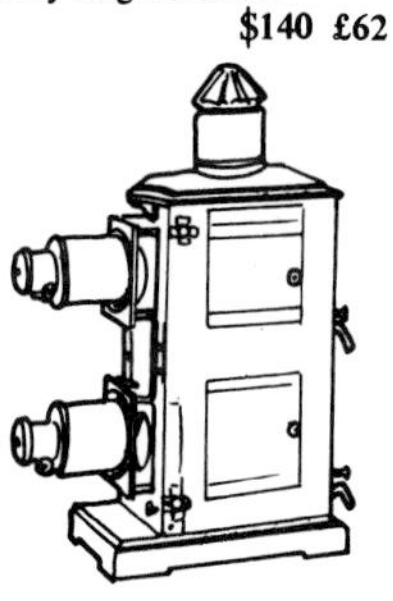

Late 19th century E. G. Wood, double lantern, 2ft.5in. high. $765 £340

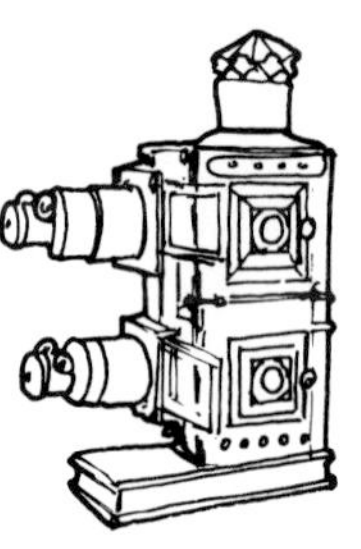

Late Victorian mahogany and brass biunial magic lantern. $800 £355

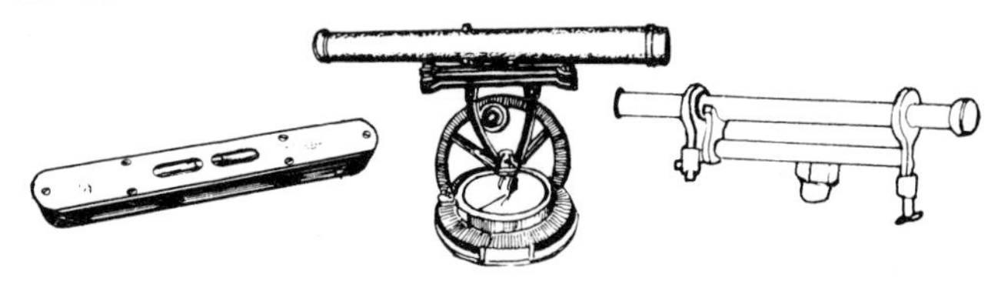

Brass and wood joiner's level by J. Buist, 1850. $35 £15

19th century surveyor's brass instrument. $250 £110

Surveyor's level in pine-wood box, circa 1840, 14ins. long. $260 £115

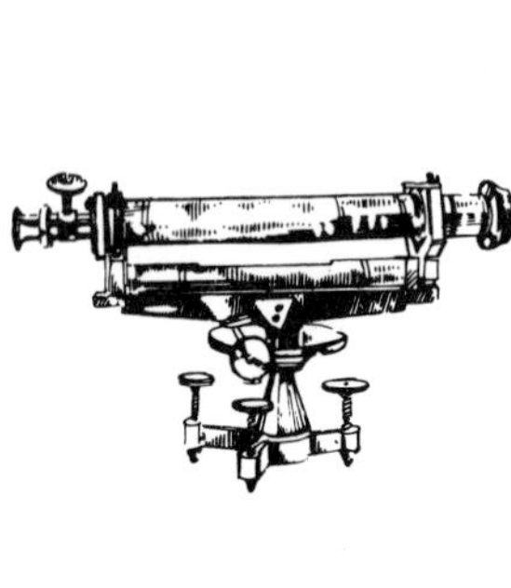

Early 20th century Troughton & Simms brass level. $270 £120

Late 19th century Elliot brass level, 1ft. 6in. long, on wooden tripod. $325 £145

A surveyor's fine sighting level in brass, by T. Cooke, York, 9in. wide. $488 £225

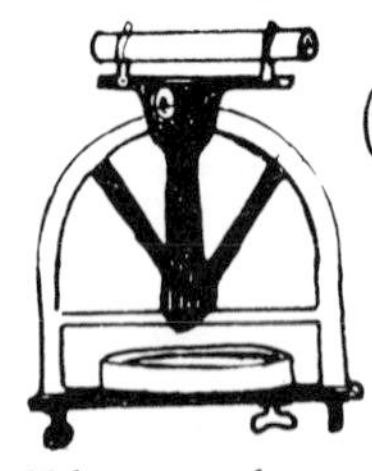

Mahogany cased surveyor's magnetic compass and sights, by W.C. Cox of Devonport. $640 £285

Mahogany cased surveyor's sighting level by Troughton & Simms, London, circa 1860, 14½in. long. $710 £315

Rare, surveyor's level by Nairn & Blunt, London, circa 1780. $1,215 £540

Early 20th century brass cash till in working order.
$180 £80

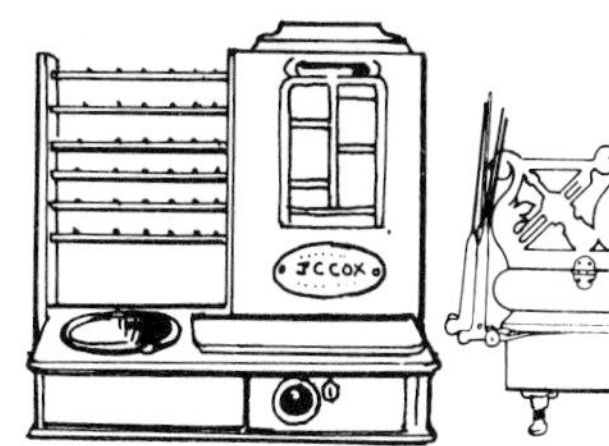

J.C. Cox patent coin changer, circa 1865-85. $225 £100

Late 19th century Padbury's patent 'Indispensable' music leaf turner, 9in. wide. $235 £105

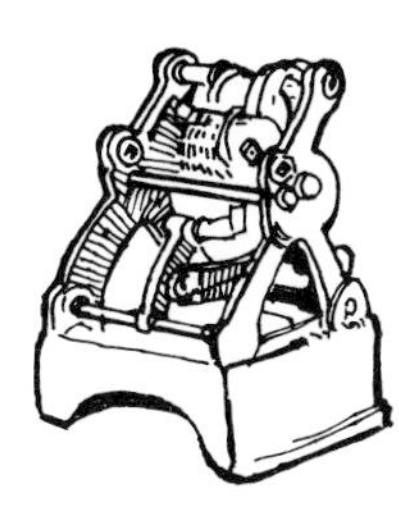

Brass and cast iron ticket dating machine, circa 1839, 8¼in. high. $325 £145

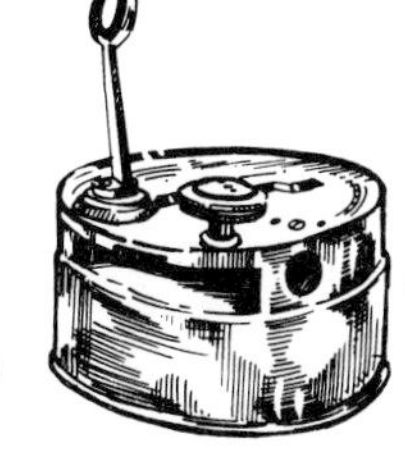

Early 19th century Adams brass box, 4in. diam.
$405 £180

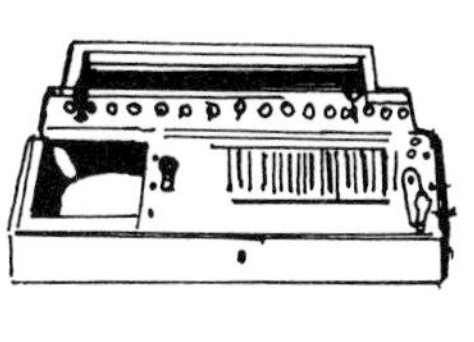

Layton's improved brass arithometer, 1ft.11½in. wide, circa 1912.
$510 £225

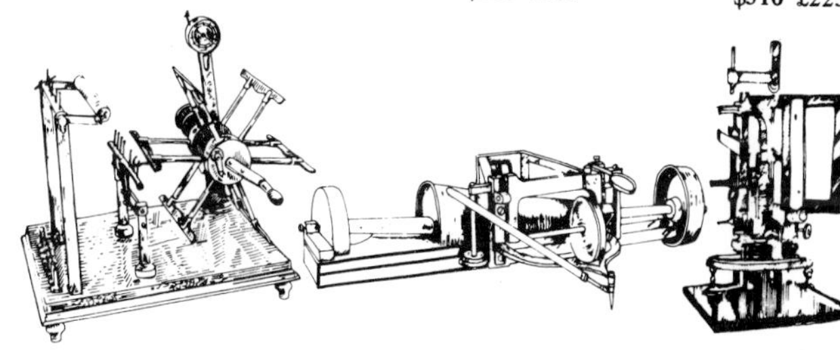

Goodbrand & Co. revolution counter on mahogany base, circa 1880, 28in. high.
$565 £250

19th century draughtsman's brass planometer by Sang, overall length 380mm. $840 £360

Rare inclinometer engraved 'Dover Charlton', Kent.
$1,350 £600

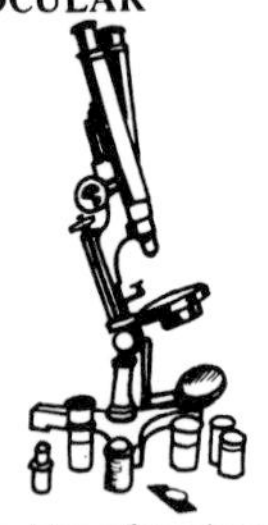

Brass binocular microscope by R. & J. Beck, 1ft.2in. high, circa 1850. $325 £145

Binocular version of the 'Watson' Edinburgh microscope. $325 £145

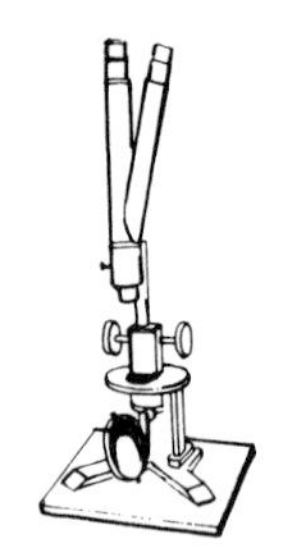

Victorian brass binocular microscope by Baker of London, circa 1850. $405 £180

Late 19th century binocular microscope with Tandem focusing to the eye pieces, 21in. high. $450 £200

Mid 19th century Pillischer brass binocular microscope, 13in. high, sold with accessories. $450 £200

Binocular microscope by Dancer, circa 1880. $530 £235

Late 19th century brass binocular microscope by A.C. Collins, 1ft.6in. high. $595 £265

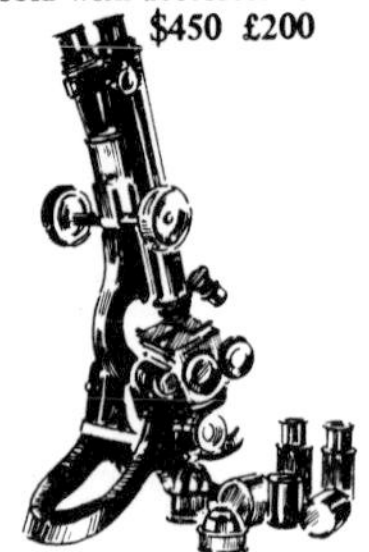

Late 19th century Swift & Sons brass binocular microscope, 1ft.5in. high. $595 £265

Late 19th century English brass binocular microscope by Henry Crouch, contained in a mahogany carrying case. $620 £275

Late 19th century R. & J. Beck folding brass binocular microscope, 1ft. 3in. high. $675 £300

Mid 19th century Negretti & Zambra brass binocular microscope, 1ft.2in. high. $675 £300

Good Baker brass binocular microscope, circa 1860, 19in. high.

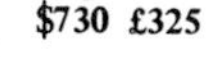

$730 £325

Late 19th century English brass binocular microscope in original case, 20in. high.
$775 £345

A Swift & Sons brass binocular microscope, 42cm., English, late 19th century.
$845 £375

A. Abrahams & Co. brass binocular microscope, 1ft. 6in. high, 1873.
$865 £385

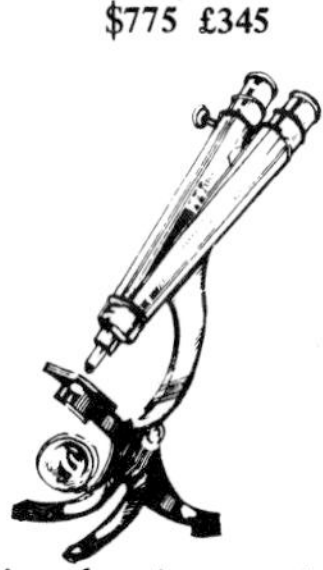

Binocular microscope in brass by Wray of London, circa 1850. $900 £400

Brass binocular microscope by Robinson of Dublin, complete with carrying case. $900 £400

Late 19th century brass microscope by R. & J. Beck. $900 £400

MICROSCOPES
BINOCULAR

A brass, compound, universal microscope, circa 1820. $1,125 £500

Smith, Beck & Beck binocular microscope, 1ft.6in. high, circa 1860.
$1,125 £500

Mid 19th century brass binocular microscope by Hugh Powell of London, 18½in. high.
$1,200 £535

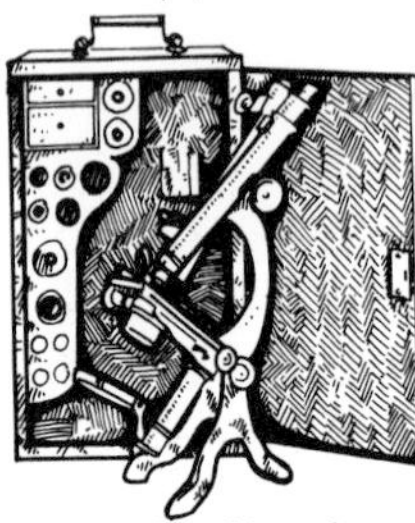

Brass Wenham binocular microscope by Armstrong of Manchester, complete with carrying case.
$1,465 £650

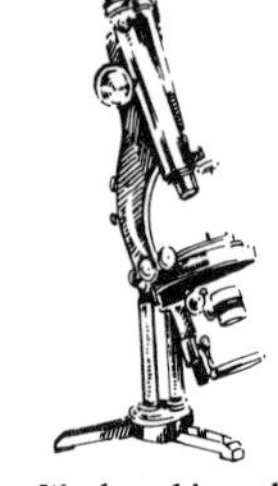

Brass Wenham binocular microscope, No. 7085, case 500mm. high.
$1,465 £650

Fine John Benjamin Dancer brass binocular microscope, 1861. $1,575 £700

Heavy brass binocular microscope, 1862, by Smith & Beck, London.
$1,755 £780

Powell & Lealand brass binocular microscope 1ft. 5in. high. $3,150 £1,400

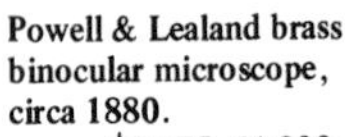

Powell & Lealand brass binocular microscope, circa 1880.
$4,275 £1,900

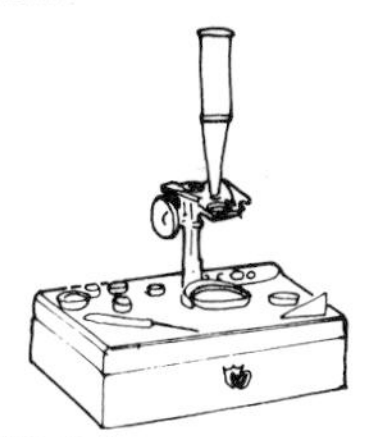

A J.P. Culls, Toms & Sutton 'Cary type' pocket microscope, signed on the pillar, which is screwed to a mahogany case, 20cm. wide, circa 1820. $450 £200

Early 29th century Cary pocket microscope, 6½in. wide. $575 £255

19th century Cary type microscope, length of box 200mm. $900 £400

CULPEPER

Early 19th century Culpeper type microscope with trade label of Wm. Harris. $675 £300

Brass microscope to an original Culpeper design of 1725. $790 £350

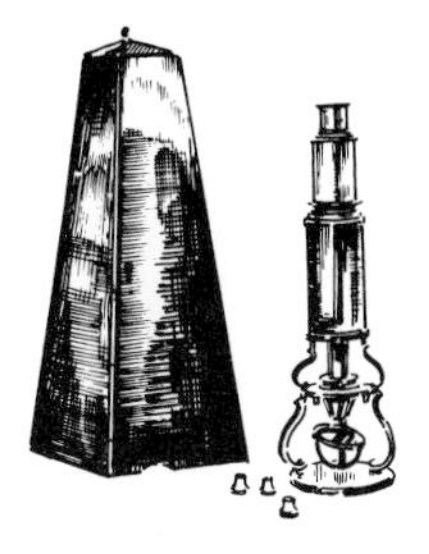

Mid 19th century Culpeper type monocular microscope, 11in. high. $865 £385

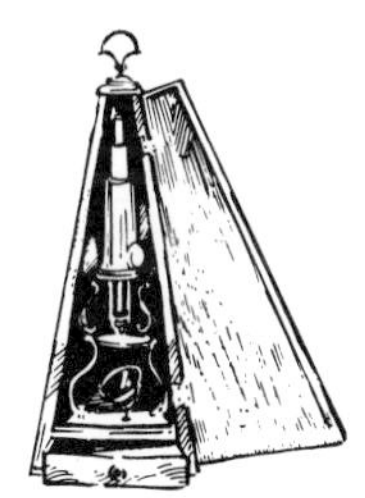

Early 19th century brass Culpeper type microscope, box 441mm. high. $1,090 £485

Early Culpeper microscope, 14in. high, with drawer of objectives and slides. $2,700 £1,200

Early 18th century Culpeper microscope, 12¾ins. high. $4,500 £2,000

MICROSCOPES
MONOCULAR

20th century iron and brass microscope, 7ins. high. $22 £10

Late 19th century brass pocket microscope by T. Rowley. $55 £25

Small 19th century brass microscope. $55 £25

Victorian brass microscope by C. Baker, London. $108 £48

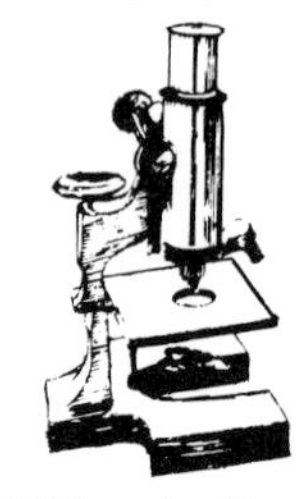

Mid 19th century French dissecting microscope by Nachet et Fils, Paris. $115 £50

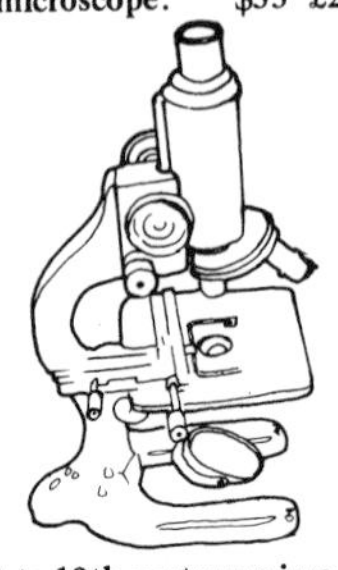

A late 19th century microscope, by W.W. Scott, in case. $125 £55

Microscope by C. Baker, London, (9184) in case. $135 £60

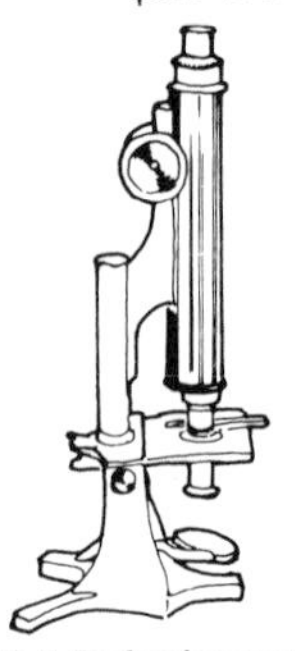

Smith & Beck microscope, in case. $170 £75

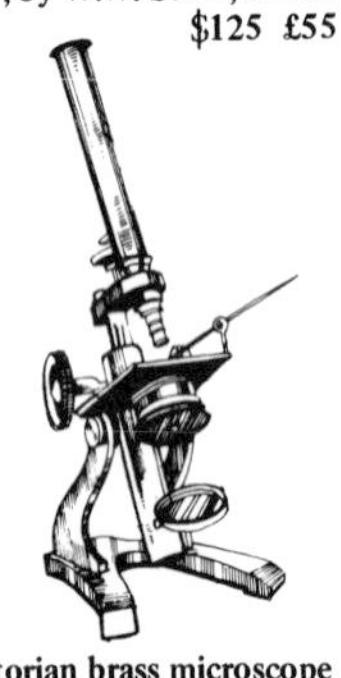

Victorian brass microscope by Baker of Holborn. $190 £85

Watson 'Fram' brass microscope, circa 1890. $200 £90

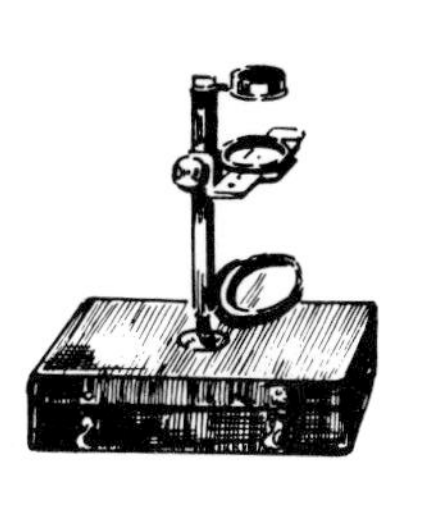

Early 19th century simple microscope with racked column, 4¼in. wide. $225 £100

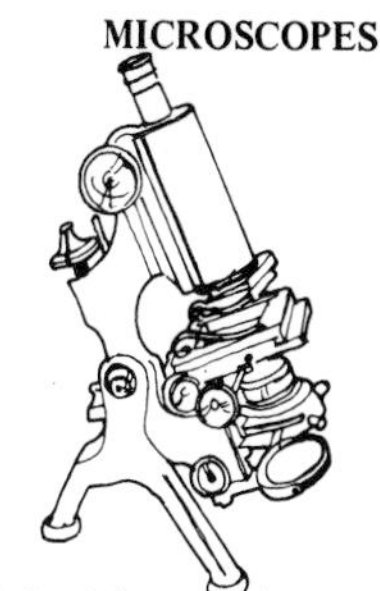

Early 20th century brass monocular microscope by W. Watson & Sons, in a rosewood case, 36cm. high. $235 £105

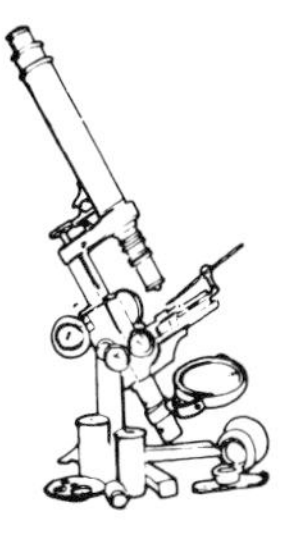

Compound microscope probably by Pillischer of Bond Street, circa 1860. $270 £120

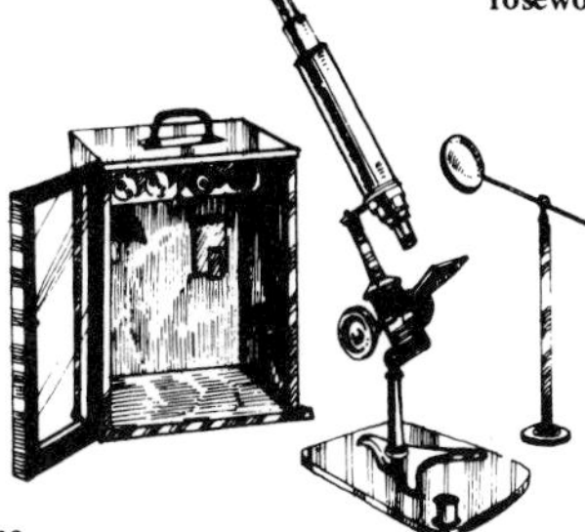

Nicely made brass microscope by Negretti & Zambra, circa 1865. $270 £120

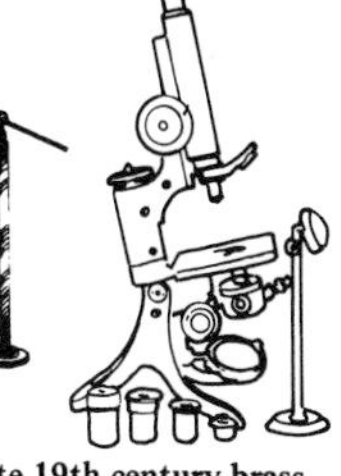

Late 19th century brass monocular microscope by Henry Crouch. $270 £120

Mahogany cased 17in. monocular brass microscope by Moginie of Finchley, London. $295 £130

Monocular compound microscope, circa 1870, 13¼in. high. $325 £145

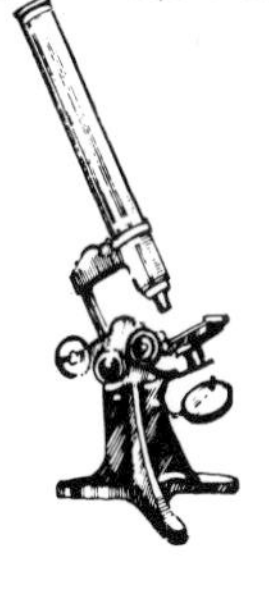

Brass microscope by John Browning, circa 1850. $340 £150

A monocular microscope, circa 1820 by D. Davis, of the original lacquered brass, 10in. high, in a mahogany case with ebony stringing, complete with all accessories. $405 £180

Powell & Lealand student monocular microscope, 1ft.6in. high, circa 1845-55. $430 £190

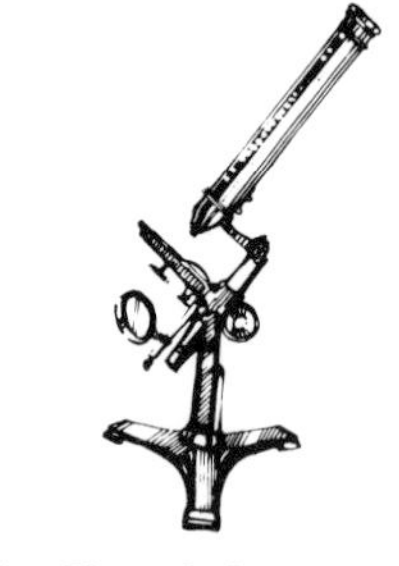

Monocular brass microscope by W. J. Salmon, 16in. high, circa 1850. $430 £190

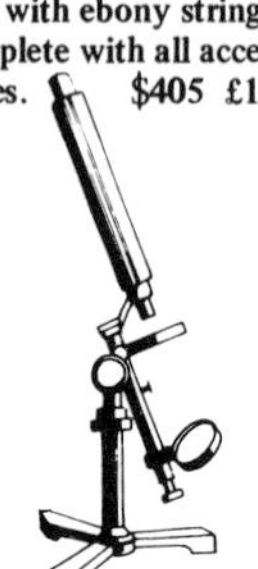

A good Victorian brass microscope by T. Cross. $450 £200

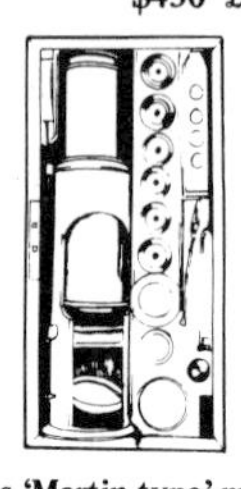

A brass 'Martin type' monocular microscope, the barrel inscribed J. Crichton, 112 Leadenhall St., London, circa 1840. $450 £200

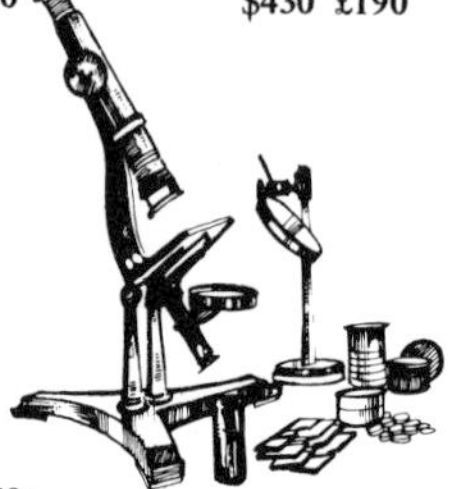

English monocular microscope by J. Benjamin Dancer, Manchester, circa 1860, 1ft.2in. high. $450 £200

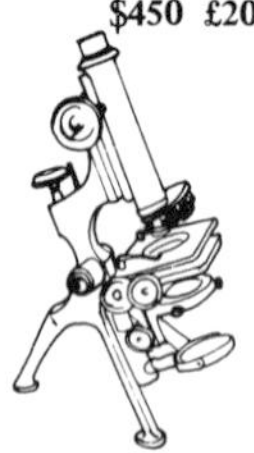

A Swift & Sons brass monocular microscope in a mahogany case, with two boxes of slides, English, early 20th century. $450 £200

Mid 19th century J.B. Dancer monocular microscope, 1ft.2in. high. $540 £240

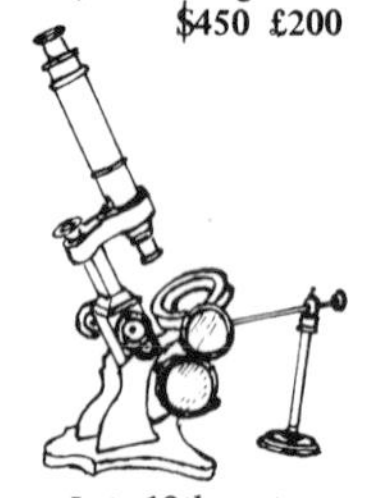

Late 19th century Baker brass monocular microscope, 15in. high. $540 £240

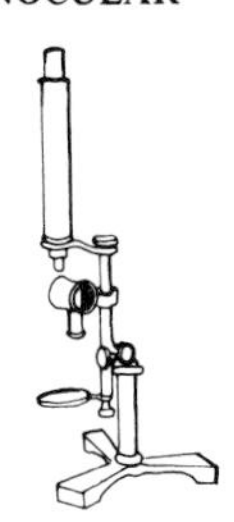

Early 19th century mahogany cased microscope by J. Cross of London, circa 1820. $585 £260

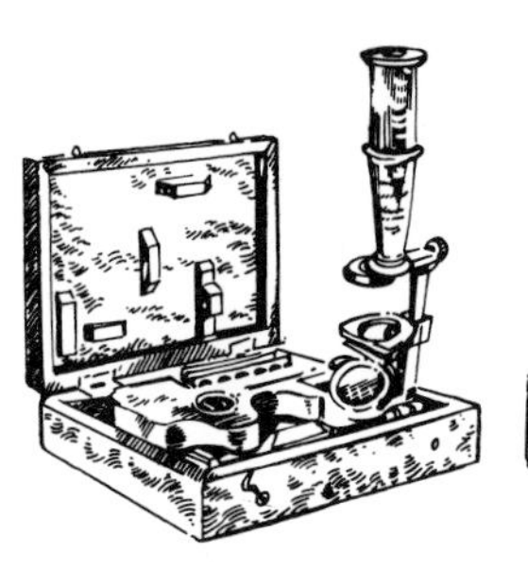

Dollands chest type brass monocular microscope, circa 1840. $700 £310

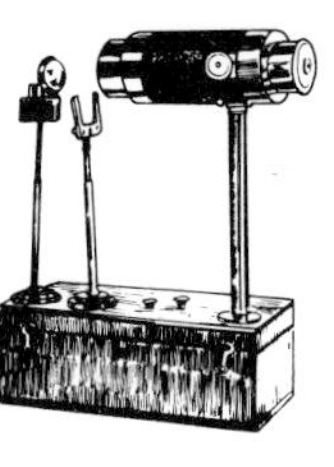

Cary brass monocular microscope, circa 1830. $700 £310

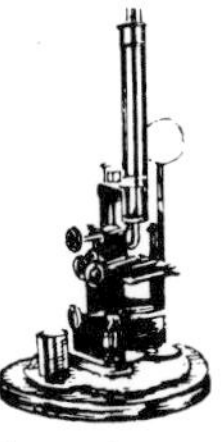

Brass microscope by Ross of London. $900 £400

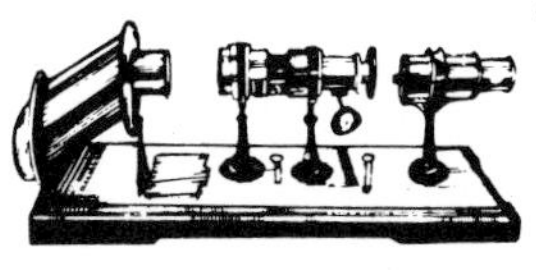

A very rare mid 19th century laboratory bench microscope, of lacquered brass and bronze, by Newton & Co., Fleet Street. $945 £420

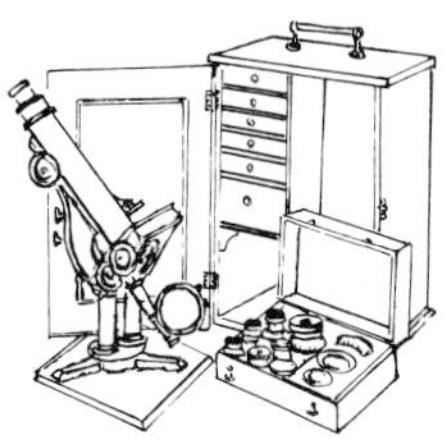

Mid 19th century English Smith & Beck brass monocular microscope, 18in. high. $945 £420

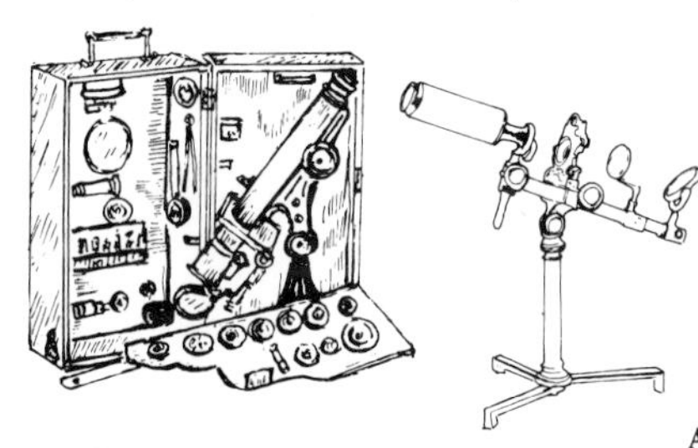

Good Smith & Beck brass monocular microscope, mid 19th century, 17in. high, with accessories. $945 £420

Georgian brass microscope by Gilbert & Sons of London. $1,070 £475

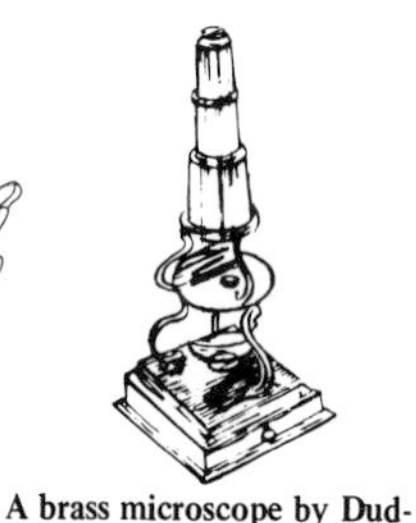

A brass microscope by Dudley Adams, The Strand, London, 1790, mounted on a mahogany box fitted with a drawer. $1,125 £500

An early 19th century lacquered brass opaque solar microscope by Wes Jones. $1,125 £500

A 19th century brass opaque solar microscope by Cary of London.
$1,125 £500

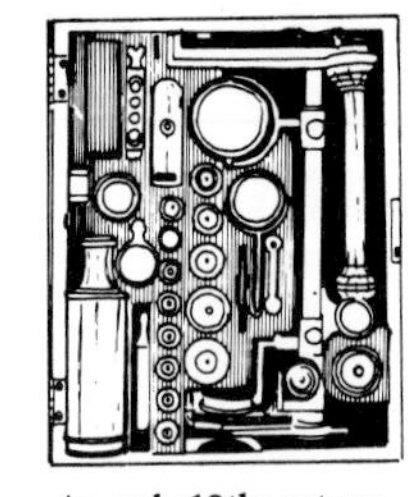

An early 19th century brass microscope by A. Abrahams of Liverpool.
$1,385 £615

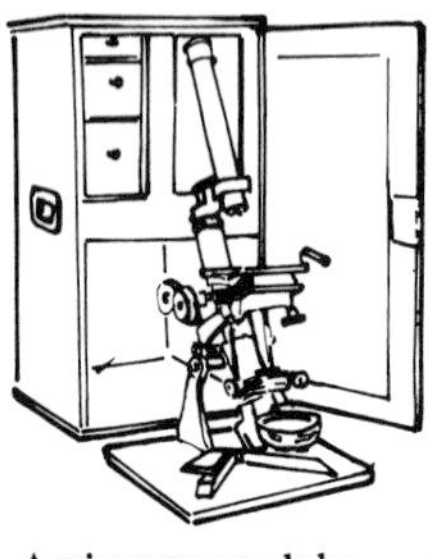

A microscope made by Powell & Lealand.
$1,385 £615

19th century brass monocular microscope by A. Ross.
$1,510 £670

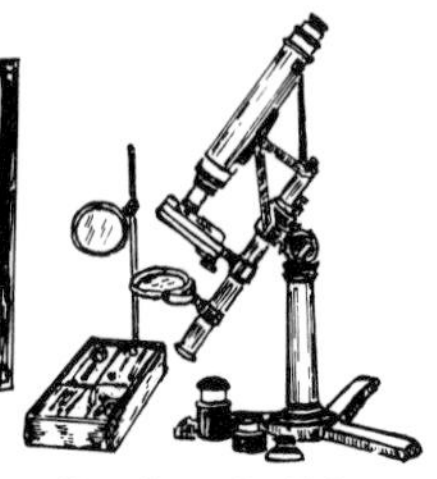

Rare James Smith brass monocular microscope, circa 1840, 17in. high, with accessories.
$1,575 £700

English microscope by Powell & Lealand, 1872.
$1,575 £700

Brass Cuff-type microscope, signed Dolland, London, 34cm. high.
$1,690 £750

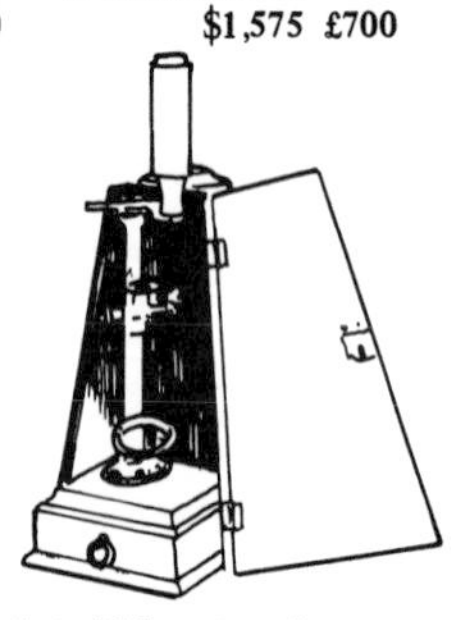

Late 18th century Jones improved type microscope.
$1,690 £750

Early 19th century brass microscope by West, Drury Lane, London. $1,755 £780

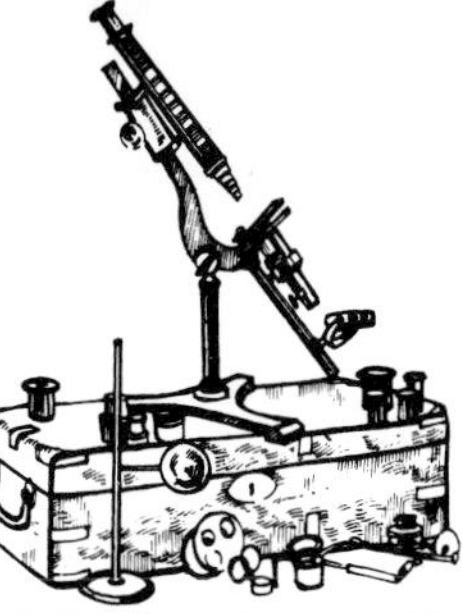

Andrew Ross microscope with accessories, circa 1839. $1,755 £780

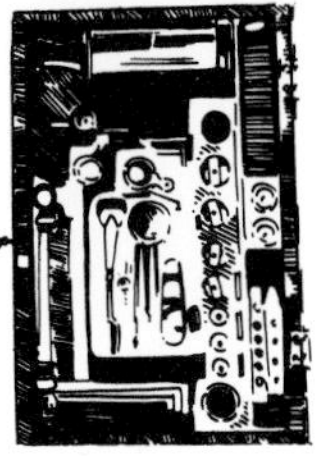

Early 19th century brass microscope by Dolland, London, in box. $1,890 £840

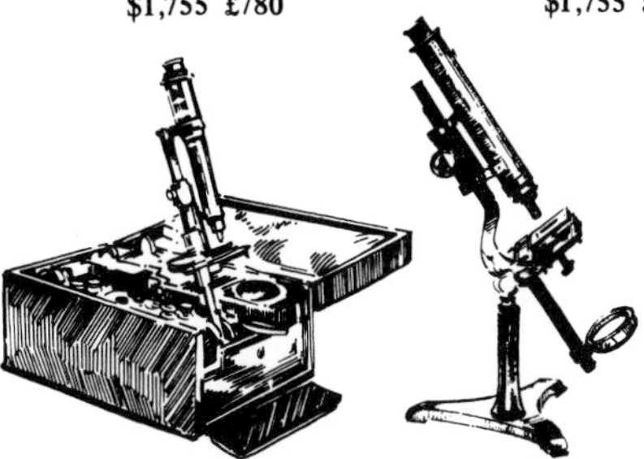

Mid 18th century chest microscope by Lincoln of London. $1,890 £840

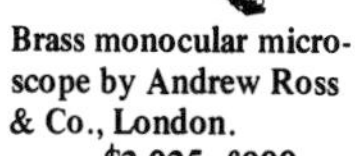

Brass monocular microscope by Andrew Ross & Co., London. $2,025 £900

19th century monocular microscope in brass by Smith & Beck, London, sold with glass slides. $2,810 £1,250

An all brass microscope by John Cuff, London, 1740. $3,375 £1,500

An important microscope by John Marshall of London, 1715. $7,875 £3,500

Magnificent microscope made by Alexis Magny, circa 1750, for Madame de Pompadour. $112,500 £50,000

Victorian metronome in an oak case. $70 £30

London-made pill rolling slab. $70 £30

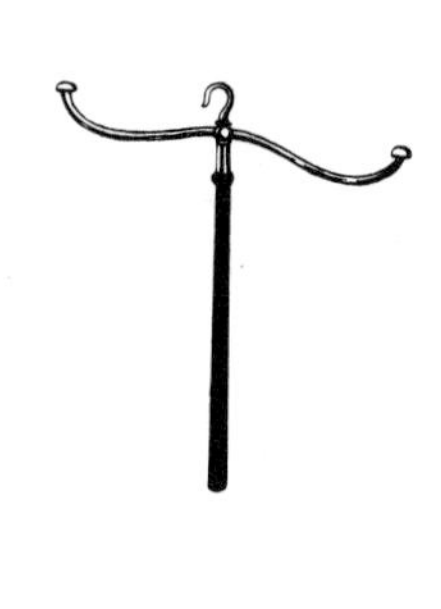

Georgian mahogany and brass wig stand. $80 £35

English spring badger trap in polished steel, in perfect working condition, circa 1760, 17¼ins. long. $80 £35

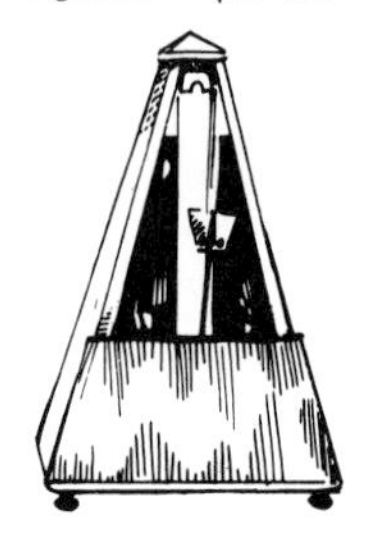

Fine pyramid walnut cased metronome 9in. high. $115 £50

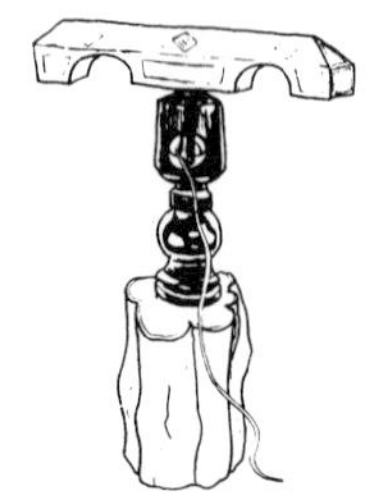

A fine and rare, Georgian, wild duck lure, being a revolving wooden bar inset with glass panels and mounted on a turned wood stand. $135 £60

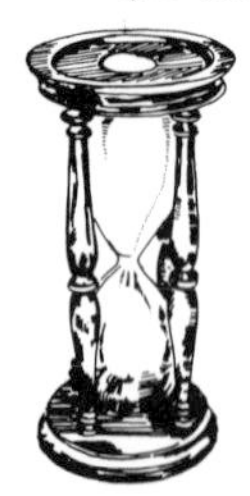

Beechwood egg timing sandglass, 19th century, 6½in. high. $155 £70

An unmarked masonic jewel, the sun gilt, circa 1800, 9cm. long. $170 £75

A very rare and interesting Imperial German Army regimental 'Casino' ivory hammer of the 74th Infantry Regiment. $170 £75

George III masonic jewel, 4½in. high, by Nicholas Cunliffe, Chester, circa 1800. $225 £100

A late 19th century French panoptical panorama, 6ins. long. $325 £145

19th century brass heliograph on an extending base. $450 £200

Early James Watt duplicating machine in mahogany box with brass rollers, circa 1825, 44.5cm. wide. $475 £200

Late 18th century set of eight tuning forks in case. $565 £250

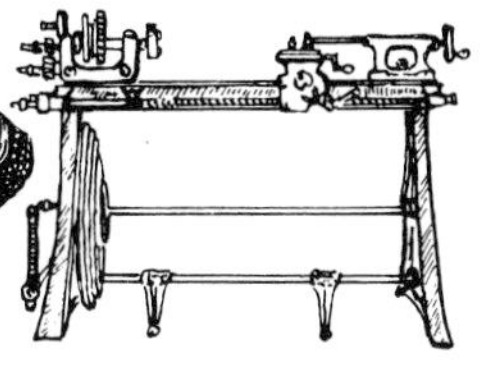

19th century G. Birch & Co. ornamental turning or brass finisher's lathe, sold with attachments. $595 £264

Mid 18th century South German or Austrian gunmaker's mainspring cramp, 5in long. $1,125 £500

Evans resuscitation apparatus, circa 1774, 1ft.6in. wide. $1,350 £600

Set of Napier's bones in boxwood, 4in. long case. $2,250 £1,000

A small 19th century mortar and brass pestle. $55 £25

Large Victorian cast iron pestle and mortar. $70 £30

A small cast iron, chemist's mortar with two handles and four decorative bands, circa 1840, 3¼ins. high. $70 £30

Brass pestle and mortar, around 1890. $80 £35

Cast iron apothecary's mortar with brass pestle, circa 1790. $90 £40

English bronze pestle and mortar, circa 1690. $340 £150

A large Continental brass mortar embossed M.D.L. XXXIV. $385 £170

Early 17th century yew wood mortar, 6¼in. high. $1,155 £525

17th century bronze mortar with the inscription 'William Boult, 1654'. $1,350 £600

English silver mounted shagreen necessaire, 24½in. high, circa 1770.
$495 £220

Rectangular pinchbeck necessaire, 2½in. wide, circa 1770. $675 £300

Gold-mounted Vernis Martin necessaire, 3½in. high, Paris 1775-1781. $730 £325

Late 19th century French opaline and gilt necessaire.
$810 £360

Gold mounted Vernis Martin Souvenir d'Amitie, Paris, 1768-1775, 3¼in. high.
$900 £400

Gold mounted Vernis Martin Souvenir d'Amitie, 3¼in. high, 1772.
$925 £410

Mother-of-pearl and wood musical necessaire, 8in. wide, circa 1820.
$945 £420

French wood musical necessaire, 9½in. wide, circa 1819-1838.
$1,080 £480

'Palais Royal' mother-of-pearl and gilt-metal musical necessaire, 5½in. wide.
$2,475 £1,100

Late 18th century ebony and brass octant with ivory scale, with green baize lined mahogany case. $450 £200

Nelson period ebony and brass octant, circa 1810. $510 £225

Early 19th century ebony octant by Richardson, London, with ivory scale, 29cm. radius, in oak case. $525 £220

Early 19th century ebony octant by H. Hemsley, with an ivory scale, 10in. radius. $565 £250

A fine late 18th century ship's captain's octant, of ebony and brass with ivory scales, by I. Steele & Son, Liverpool, complete with original oak carrying case. $730 £325

A very fine ship's captain's octant, made of ebony and brass with ivory scales, by I. Steele & Son, Liverpool, circa 1790. $790 £350

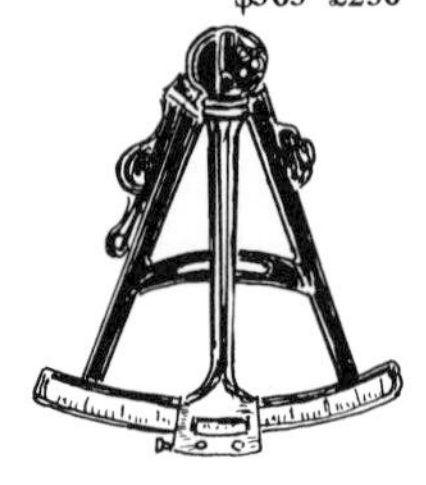

Ebony and brass Naval Officer's octant with ivory scales, by Crichton, London, circa 1825. $865 £385

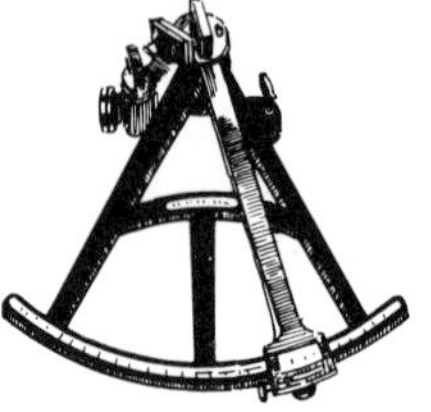

Early 19th century Cary octant with ebony frame, 11½in. radius. $1,035 £460

An ebony octant with ivory scales by Jones, Gray and Keen of Liverpool, circa 1780. $1,240 £550

Victorian mother-of-pearl and brass opera glasses. \$45 £20

French blue lacquered metal, opera glasses, circa 1900. \$45 £20

Pair of 19th century silver cased opera glasses. \$55 £25

Victorian lorgnette style ivory opera glasses. \$80 £35

Georgian silver lorgnette with tortoiseshell case. \$80 £35

Victorian tortoiseshell handled lorgnette. \$160 £70

Mother-of-pearl and decorative enamel opera glasses. \$175 £78

George III silver opera glass, signed Dolland, London, circa 1790, 2¾ins. high, with case. \$900 £400

Pair of gold opera glasses, the body encrusted with diamonds. \$12,375 £5,500

ORRERYS

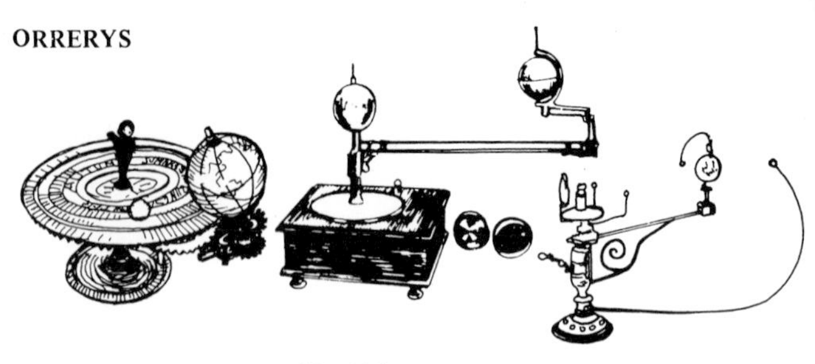

Late 19th century Salter Parkes and Hadley's patent orrery, 10in. diam.
$945 £420

Fine 19th century French enamel and brass orrery on a walnut base. $1,070 £475

Unusual late 19th century table orrery, 550mm. high.
$1,395 £620

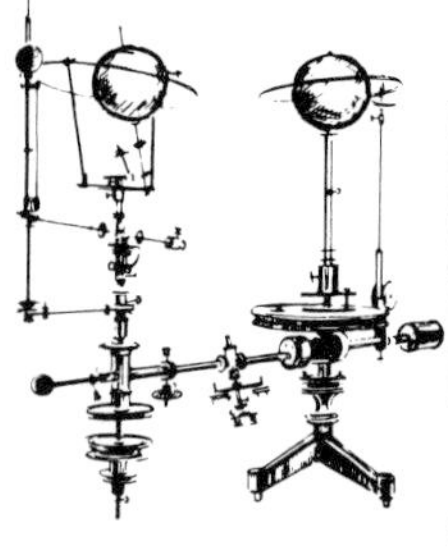

Early 20th century orrery, 3ft.5in. wide, with sectioned diagram.
$1,520 £675

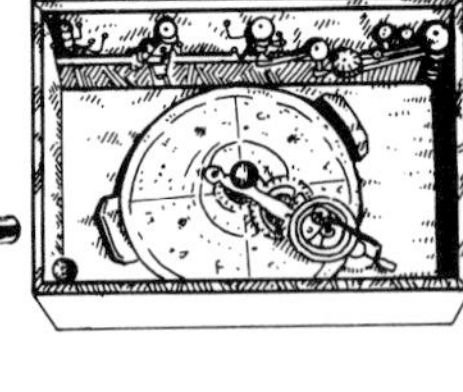

19th century portable orrery, the celestial table with scales for the months of the year.
$1,745 £775

19th century French orrery by Delmarche, Paris, brass mounted on mahogany stage.
$2,140 £950

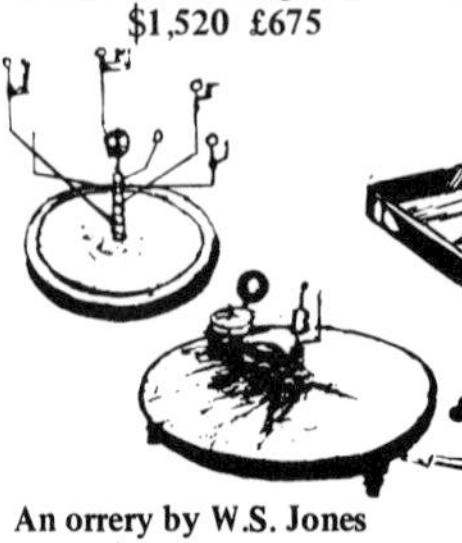

An orrery by W.S. Jones of London, 1794, 13in. diam. base with interchangeable planetarium.
$2,925 £1,300

19th century portable orrery by W. Jones, 195mm. diameter.
$3,600 £1,600

Fine English lacquered brass orrery by Thos. Harris & Son, London.
$6,410 £2,850

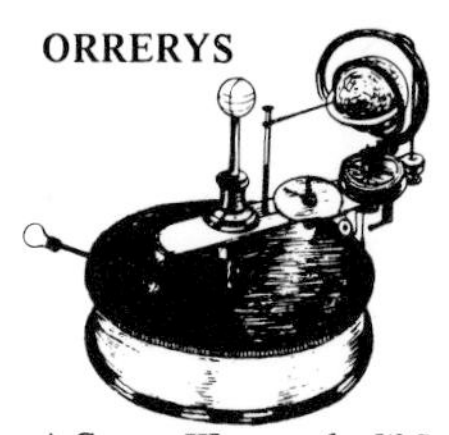

A George III orrery by W.S. Jones of London, In brass and with ivory and painted solar bodies, the terrestrial globe is dated 1800, and is contained in a mahogany case. $7,200 £3,200

A 'New Planetarium' by Benjamin Martin of London, circa 1770. $11,250 £5,000

The original and historically important orrery made by John Rawley and named after his patron the 4th Earl of Orrery. $78,750 £35,000

PANTOGRAPHS

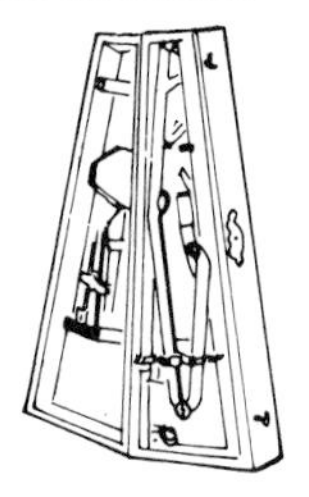

Brass pantograph by W. & S. Jones, London. $180 £80

Mahogany cased brass pantograph by 'Cole, London', 25¾in. long, circa 1780. $440 £195

Brass pantograph by 'Silberrad, London', circa 1810. $620 £275

PENKNIVES

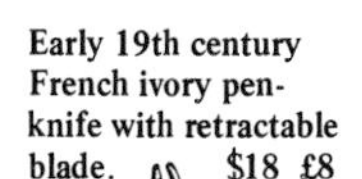

Early 19th century French ivory penknife with retractable blade. $18 £8

Silver knife, Sheffield 1856 with an ebonised handle. $35 £15

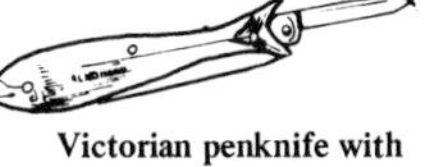

Victorian penknife with ivory fish handle. $36 £16

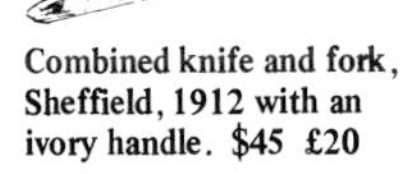

Combined knife and fork, Sheffield, 1912 with an ivory handle. $45 £20

French ivory pen machine, circa 1840. $70 £30

Mid 19th century French mother-of-pearl handled desk knife. $70 £30

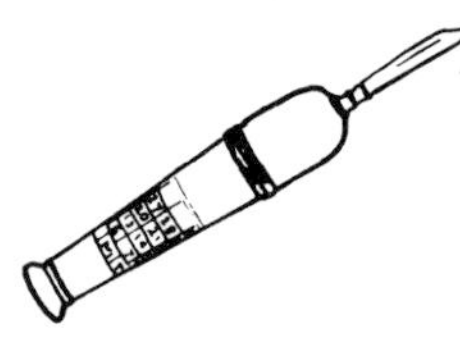

Sheffield made ivory penknife with perpetual calendar. $115 £50

Circular silver pocket almanac by Loos, 1805, 44mm. diam. $765 £340

A silver perpetual almanac by Boardman, 1709, 3in. diam. $775 £345

German silver perpetual calendar, 58mm. diam. $855 £380

Red and black perpetual calendar mantel clock, circa 1860, 17½in. high. $1,125 £500

Early 18th century German silver gilt trefoil shaped combined perpetual calendar and spice box, 62mm. long. $2,475 £1,100

PLANES

19th century beechwood coffin-shaped smoothing plane, 8½in. long, by Charles & Co. $55 £25

Beechwood and brass plough plane by J. Mosele, Bloomsbury, London. $55 £25

19th century beechwood smoothing plane, 7¼in. long. $65 £30

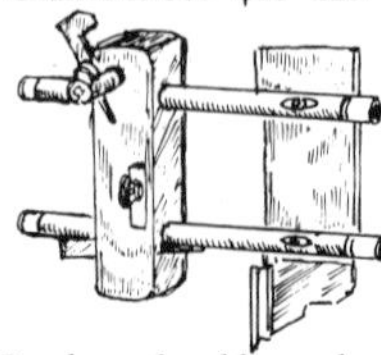

Beechwood and brass plough plane by W. Greenslade, Bristol. $85 £40

Steel and mahogany shoulder plane. $108 £48

Rare circular plane by Stanley Rule & Level Co., 1877, 10¼in. long. $175 £80

Late 19th century metal printing stamp with an ivory handle and floral decoration. $22 £10

An improved Albion Printing Press by W. & J. Figgins, London, No. 1187, 1874. $530 £235

Columbia Printing Press, with distinctive golden eagle crest. $700 £310

PROJECTORS

Candle lit cine projector complete with films. $125 £55

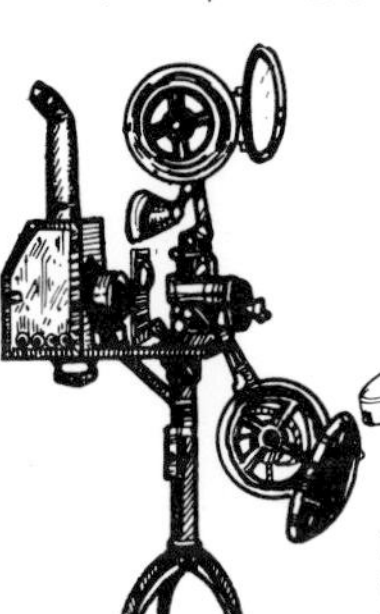

An early cinematograph projector. $585 £260

Kinora moving picture viewer with 24 reels, circa 1905, 18½in. long. $740 £340

PROTRACTORS

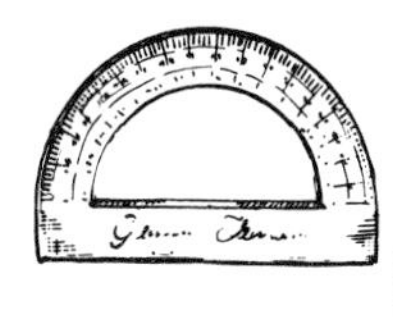

Brass protractor by Godelar of Paris, 4in. diam. $70 £30

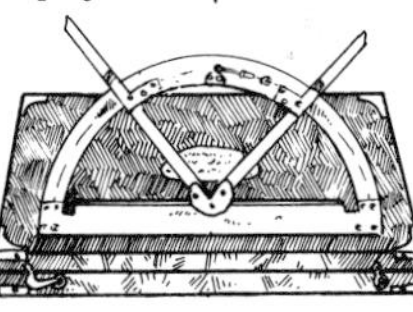

Early 19th century brass protractor by Gilbert of London. $700 £310

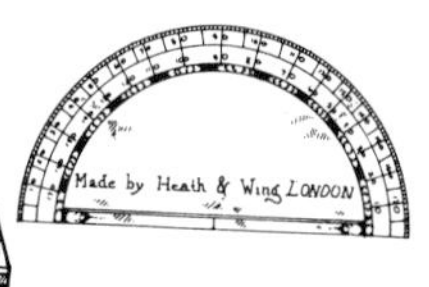

18th century silver protractor by Heath & Wing of London. $700 £310

QUADRANTS

16th century Italian gilt metal quadrant. $4,345 £1,930

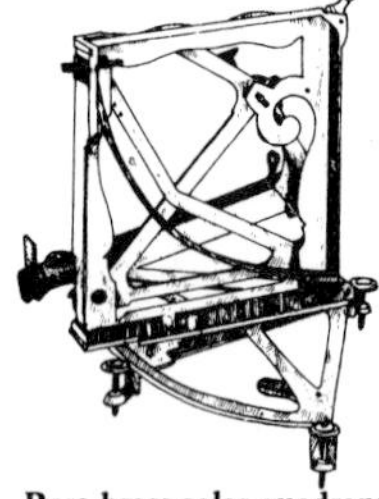

Rare brass solar quadrant by Duboia, Paris, circa 1770, in original box. $5,060 £2,200

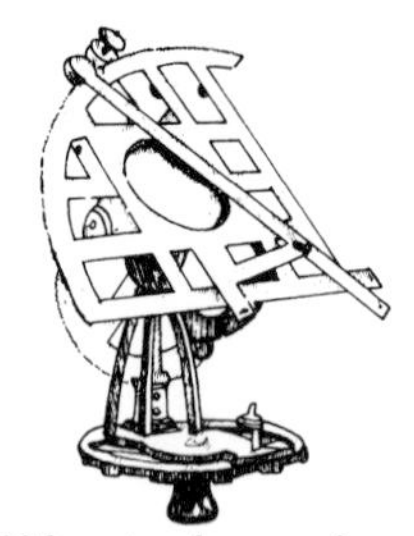

18th century brass quadrant by J. Sisson, London, 233mm. radius, incomplete. $6,900 £3,000

RADIOS

English Celestian radio speaker in a mahogany cabinet. $70 £30

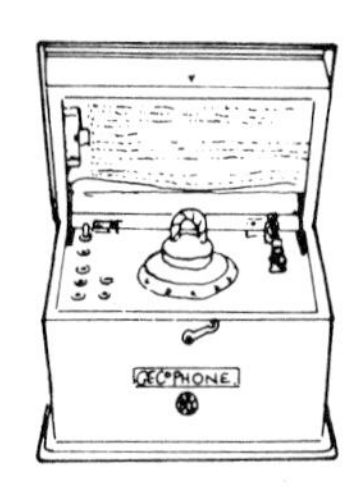

Gecophone (B.B.C.) in mahogany case. $70 £30

An English amplion radio speaker in domed wooden case. $80 £35

Single valve radio, circa 1925, with original valve and tuning coil. $160 £70

A rare American National microphone dancer, 1ft. 1in. high, circa 1935. $360 £160

Emor chromium plated radio set, circa 1950, 42½in. high. $430 £190

RULES

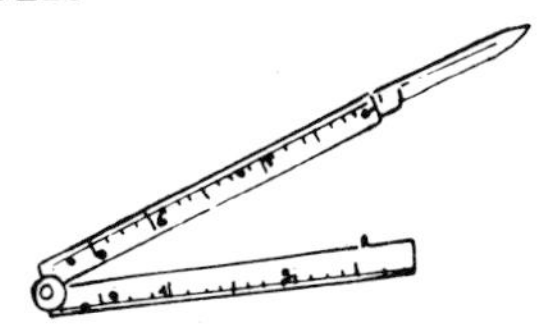

Folding ivory ruler with pen blade. $55 £25

Ebony and brass parallel rule, circa 1820. $60 £27

Fuller's rule and calculator with brass attachment. $135 £60

Rare lignum vitae and brass folding parallel rule and sighting instrument, circa 1900, 12in. long. $170 £75

English lacquered brass circular slide rule, 4in. diam., by Parsons, London. $225 £100

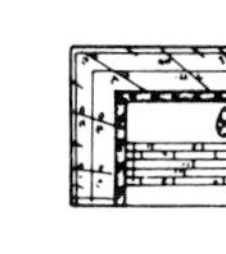

Early 18th century brass rule by Thomas Wright. $900 £400

SAWS

19th century butcher's bone saw, 23in. long. $100 £45

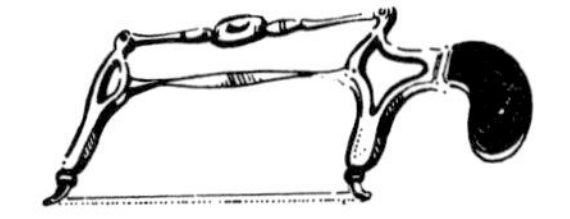

Dr. Butcher's saw by Coxeter, circa 1860. $340 £150

Edwardian brass letter scales. $35 £15

Polished steelyard sack scales, complete with pear-shaped weight, circa 1790, 25ins. long. $80 £35

Victorian kitchen scales with brass pans and cast iron stand. $100 £45

Eastman Studio scale by Kodak, New York, 9in. wide. $100 £45

Oak based brass postal scales, circa 1870, 7½in. wide. $108 £48

A set of brass letter scales with weights. $135 £60

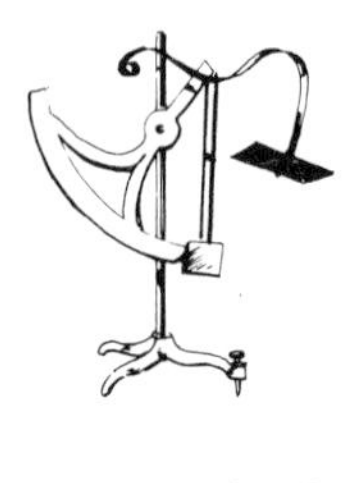

19th century Salter spring balance. $135 £60

Unusual brass and steel, gold scales by E. Levrig & Co. of London, 16ins. high, circa 1850. $170 £75

Set of 19th century brass banker's scales. $225 £100

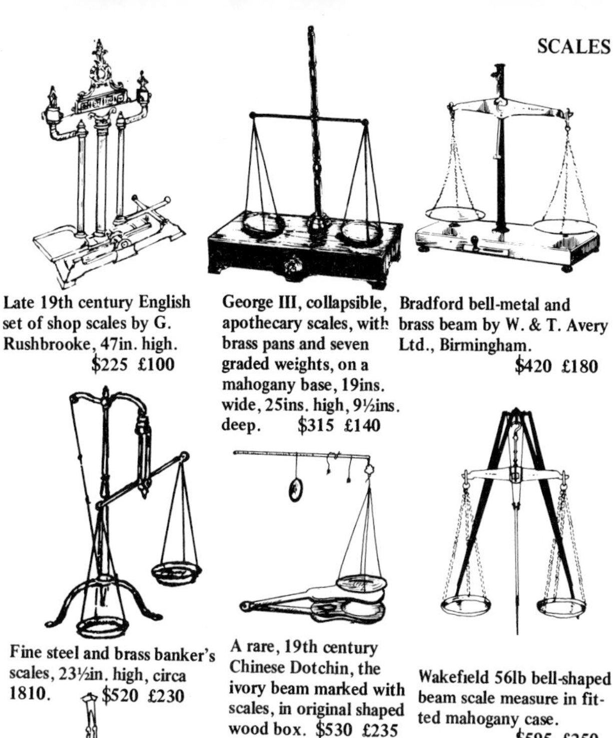

Late 19th century English set of shop scales by G. Rushbrooke, 47in. high.
$225 £100

George III, collapsible, apothecary scales, with brass pans and seven graded weights, on a mahogany base, 19ins. wide, 25ins. high, 9½ins. deep. $315 £140

Bradford bell-metal and brass beam by W. & T. Avery Ltd., Birmingham.
$420 £180

Fine steel and brass banker's scales, 23½in. high, circa 1810. $520 £230

A rare, 19th century Chinese Dotchin, the ivory beam marked with scales, in original shaped wood box. $530 £235

Wakefield 56lb bell-shaped beam scale measure in fitted mahogany case.
$595 £250

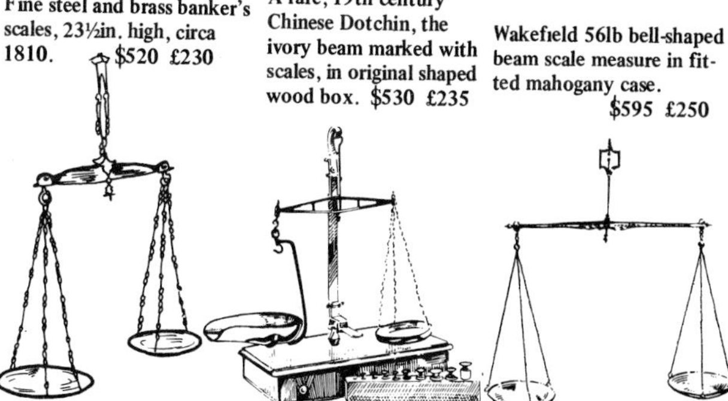

A 19th century brass balance stamped on the crossbar, F. Noriega, Madrid.
$700 £310

Set of early 19th century grocer's brass scales by Avery, with ten brass weights.
$745 £330

A pair of George III scales, London, 1793, 8in. wide.
$765 £340

Unusual jockey's scale, circa 1880, with mahogany seat. $550 £250

Brass and mahogany jockey's scale, with hide seat, circa 1850. $850 £380

Late 19th century English jockey's scale by Young. $850 £380

Victorian jockey's scales with a buttoned hide seat and bobbin legs. $1,240 £550

Mid 19th century, mahogany, leather-upholstered weighing chair, by T. Avery of Birmingham. $1,240 £550

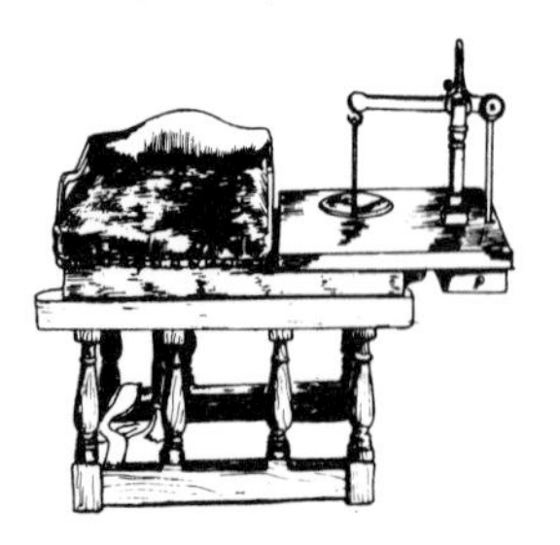

19th century English mahogany jockey's scale, the rectangular support on turned baluster shaped legs. $1,350 £600

Wanzer lock-stitch sewing machine on marble base. $55 £25

Small French sewing machine, 8in. high. $125 £55

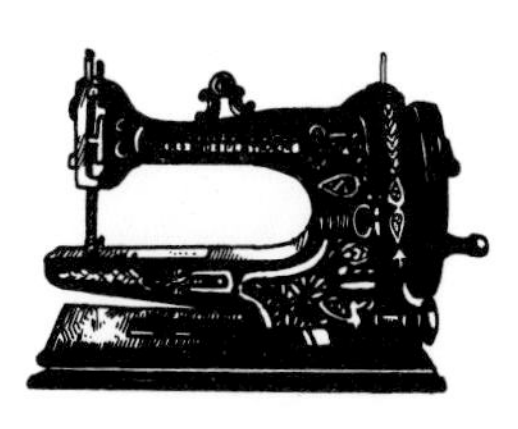

An English arm-and-platform sewing machine, dated 1877. $250 £110

Sewing machine by Wheeler & Wilson, circa 1854. $580 £260

Britannia Wheeler & Wilson lock-stitch type sewing machine with cast iron treadle table. $1,220 £500

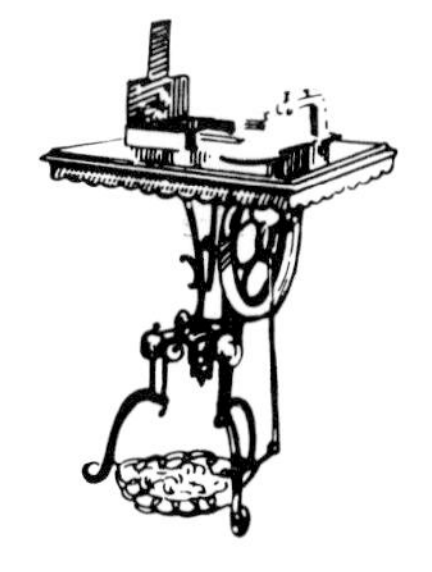

Late 19th century Wheeler & Wilson type Britannia sewing machine. $1,495 £650

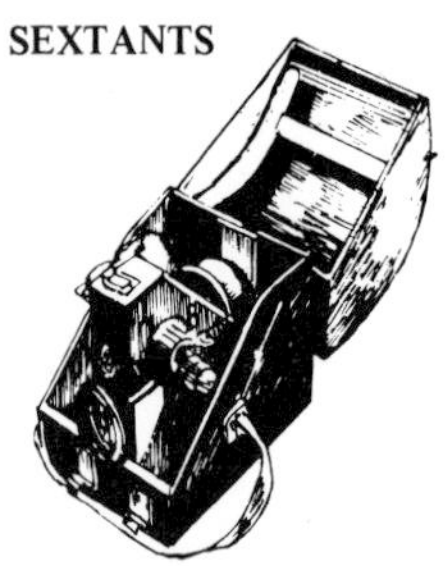

A bubble sextant, marked IX A, with an element, in case. $70 £30

Brass box sextant by Elliot Bros., in drum-shaped case, 7.5cm. high. $165 £90

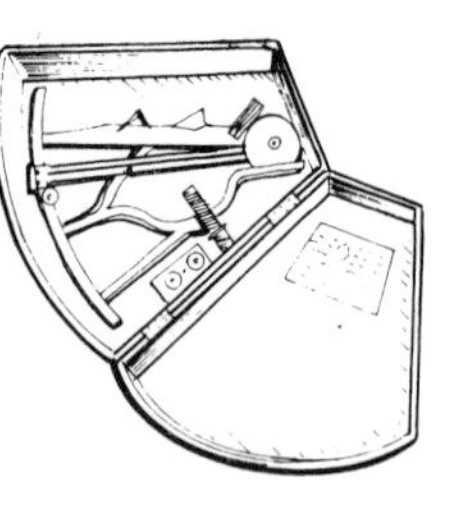

Mahogany cased brass sextant with a platinum scale, by Harrison of Hull. $295 £130

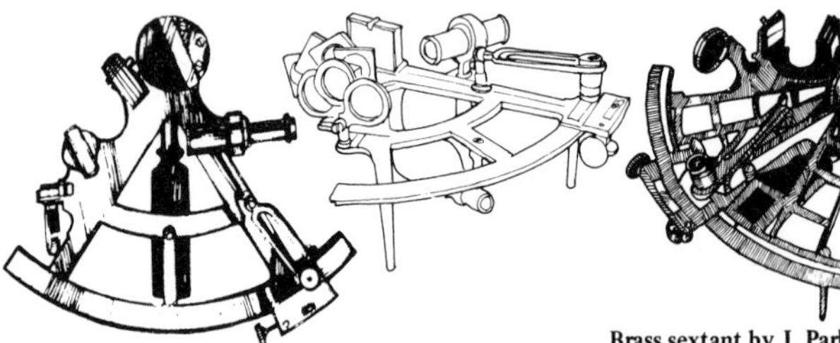

Early 19th century brass sextant by Cary of London, 8½ins. radius. $295 £130

Early 20th century Butler's brass sextant with two sets of filters. $325 £145

Brass sextant by J. Parkes & Son, the silvered scales with magnifier and Vernier adjustment. $325 £145

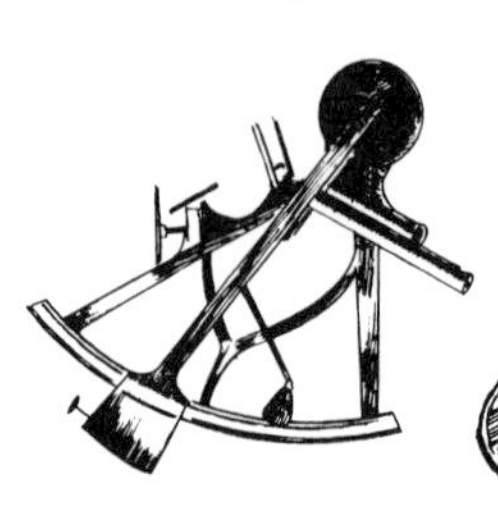

Mahogany cased brass sextant with platinum scales, by Harrison of Hull. $350 £155

Pocket sextant by Gilbert and Gilkerson, London, circa 1810. $405 £180

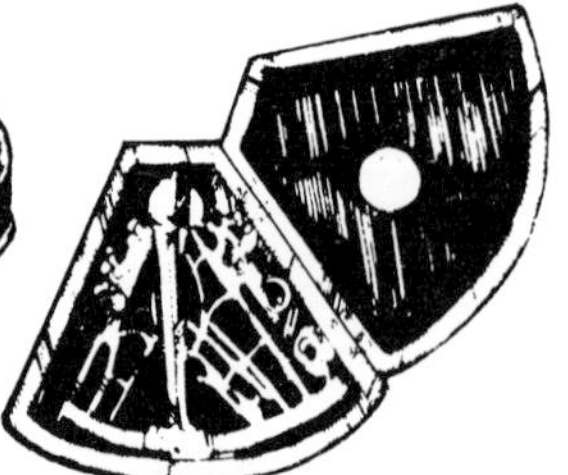

19th century cast brass sextant, in original mahogany case, by Simpson & Roberts, Liverpool, circa 1840. $440 £195

Vernier sextant by Cary, London, in fitted case.
$450 £200

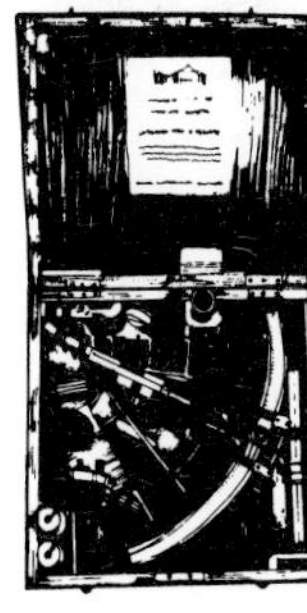

A late Victorian, metal framed sextant in its original fitted, lined, wooden case, with four spare eye pieces, by Army & Navy $475 £210

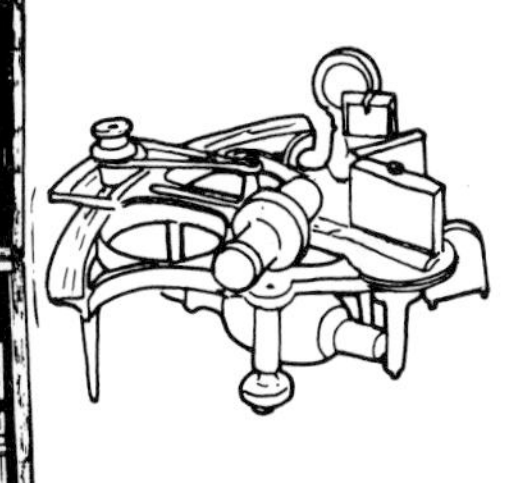

Late 19th century Andrew Christie, brass sextant, 6in. radius.$475 £210

Early 20th century Heath & Co. brass sextant, 7in. radius. $475 £210

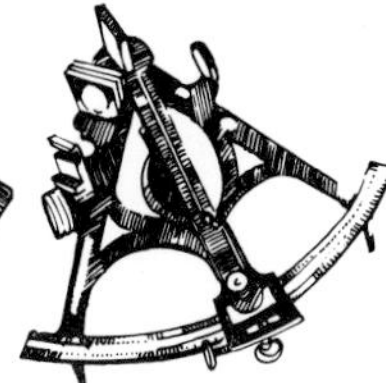

Early 19th century sextant with brass frame, signed Langford & Son, Bristol.
$495 £220

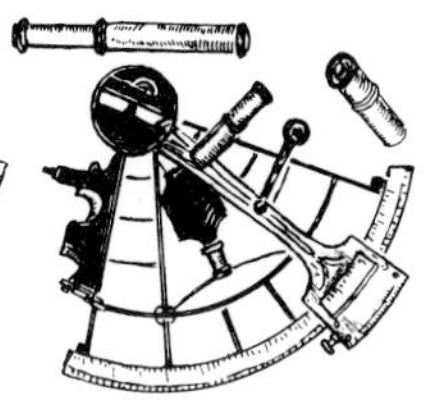

Late 19th century English brass sextant, 7½in. radius. $520 £230

Naval officer's sextant by 'A. Willings & Co.' circa 1840. $520 £230

Bronze sextant by Hayes Brothers, Cardiff, in mahogany case, circa 1850. $520 £230

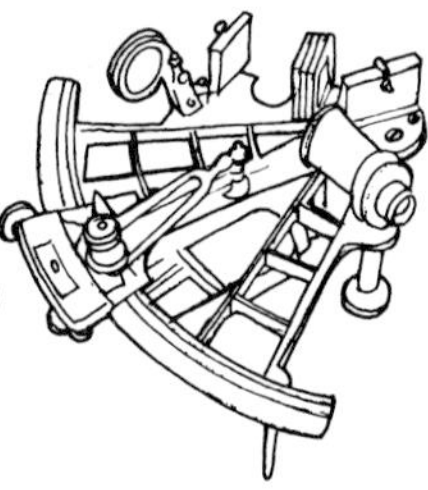

19th century brass sextant by A.F. Parkes & Sons. $530 £235

SEXTANTS

Late 19th century sextant scale signed Emanuel, Southhampton. $540 £240

A very fine 19th century sextant, by Lilley & Son, London, with platinum scale and rosewood sighting handle, in original mahogany box. $610 £270

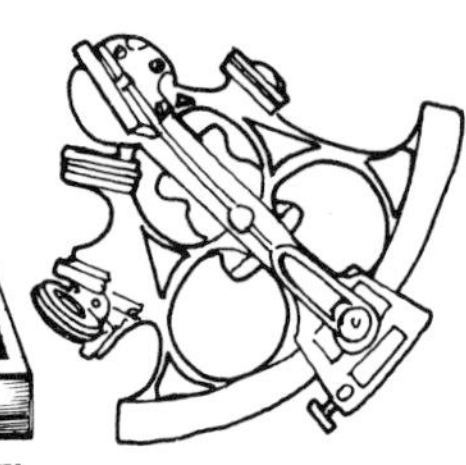

An early 20th century Cox and Coombs oxidised brass sextant with two sets of filters. $585 £260

A fine sextant by Lilley & Son, London, with platinum scale and rosewood handle, circa 1860. $585 £260

Naval Officer's brass sextant by McGregor & Co., circa 1840. $730 £325

Mid 19th century Naval officer's sextant in black bronze $790 £350

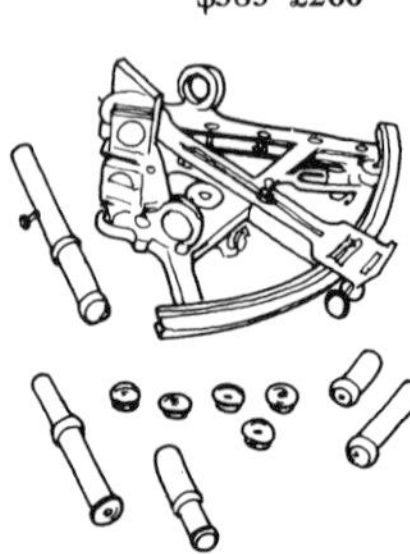

A brass sextant by Troughton & Simms. $900 £400

Captain George's R.N. patent double sextant by Elliott Bros., London, 12.5cm. diam. $810 £440

Rare mahogany cased brass gravity sextant, circa 1865. $1,180 £525

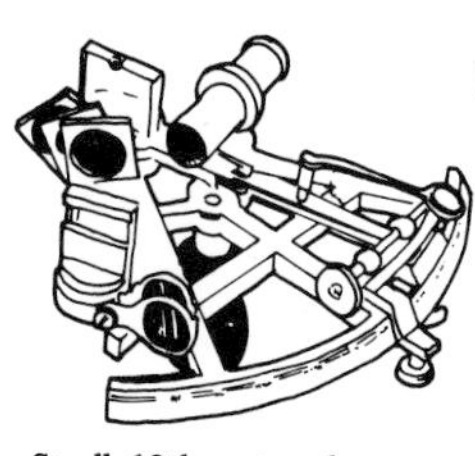

Small, 19th century brass sextant by Cary, 130mm. radius. $1,800 £800

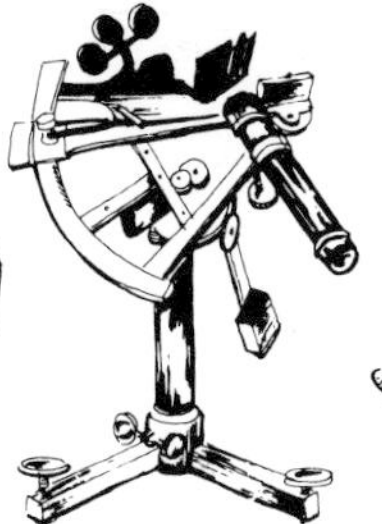

Magnificent lacquered brass 19th century pillar sextant by Cary of London. $2,250 £1,000

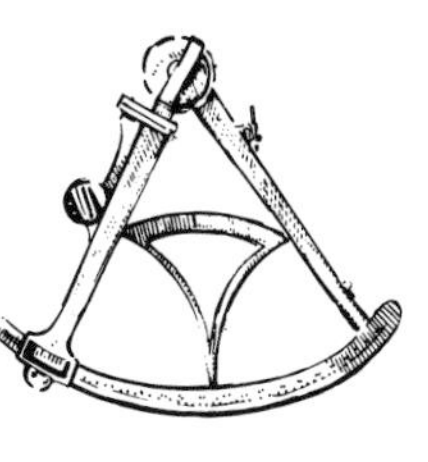

Rare brass sextant by George Adams, circa 1760, 152mm. radius. $3,150 £1,400

SHIPS BINNACLES

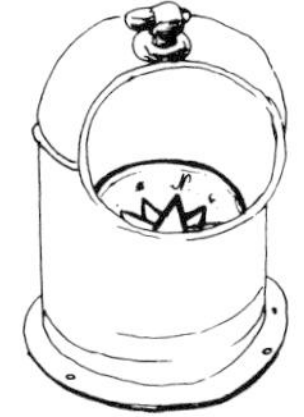

Victorian ship's compass in brass binnacle, 11in. high. $135 £60

French, ship's brass compass binnacle, by E.D. Anne, Bordeaux, circa 1870, 11in. high. $225 £100

Model of a ship's binnacle, circa 1920. $430 £190

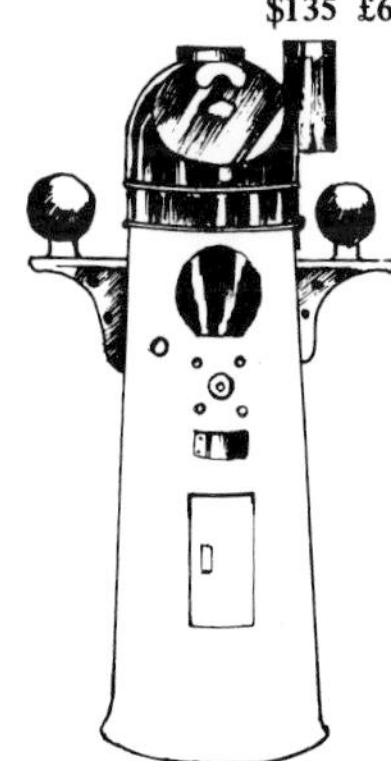

19th century mahogany and brass binnacle. $565 £250

Ship's binnacle by Hughes & Son, London, 125cm. high. $865 £385

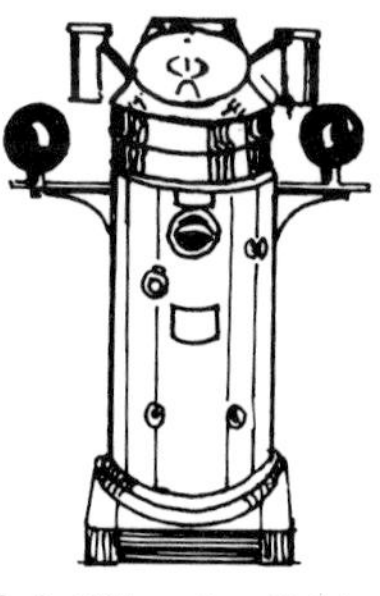

Early 20th century Kelvin, Bottomley & Baird Ltd., brass binnacle compass, 4ft. 4in. high. $1,080 £480

19th century brass ship's propeller. $55 £25

19th century copper ship's lamp. $90 £40

A ship's brass and copper binnacle cover. $90 £40

Sailing ship's watch bell, made of bell founders' brass, complete with heavy clapper and hand rope, 9½in. diam., 13¼in. high. $160 £70

Captain's, bronze, boat tiller-arm engraved 'Grecian 1812 Baltimore'. $160 £70

Small 19th century bronze model signal cannon. $180 £80

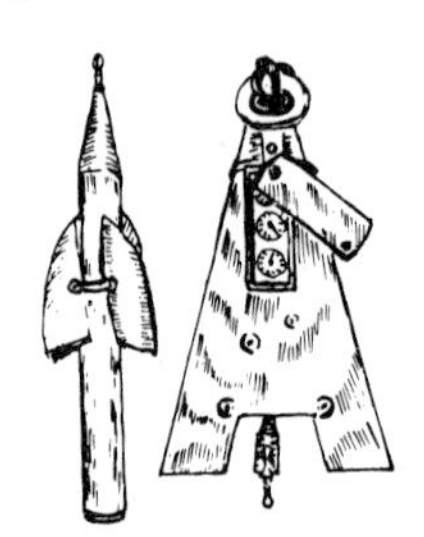

Rare early 19th century Edward Massey brass ship's log and spinner. $295 £130

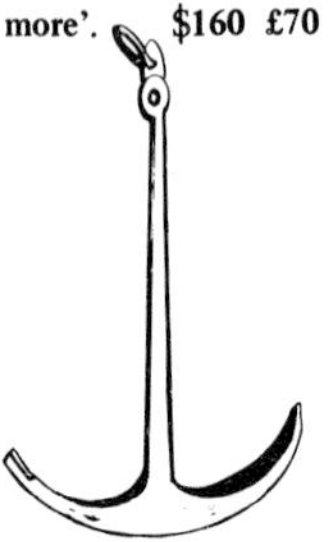

Heavy 19th century bronze anchor, 3ft. 2ins. long. $350 £155

Early mariner's instrument made in boxwood. $4,500 £2,000

SHIPS TELEGRAPH

19th century Norwegian ship's telegraph by Hynnes Maskin in Forretning, Trondheim, 43ins. high. $395 £175

Late 19th century, J.W. Ray & Co., brass ship's telegraph, 3ft.2in. high. $510 £225

SHIPS TELEGRAPH

Ship's Bridge Telegraph in brass, signed 'Bloc-tube Controls', 105cm. high. $640 £285

SHIPS WHEELS

Mahogany ship's wheel with brass fittings, 24ins. diam.. $280 £125

A fine brass-banded teak ship's wheel standing on a circular brass-banded teak column. $340 £150

19th century oak ship's wheel, 37ins. diam. $340 £150

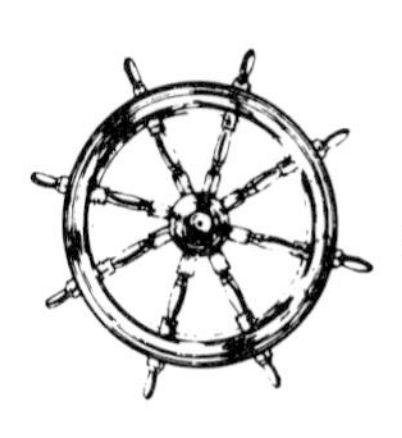

19th century brass rimmed teak Clipper ship's wheel, 41ins. diameter. $405 £180

Ship's wheel with brass fittings and inscription 'John Haste & Co., Greenock', 36½ins. diam. $510 £225

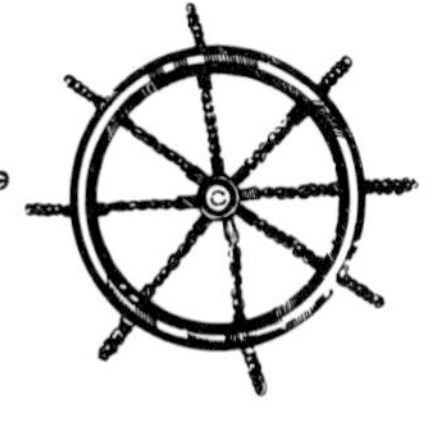

Large teakwood and brass ship's wheel, 55in. diam. $1,035 £460

SPECTROSCOPES

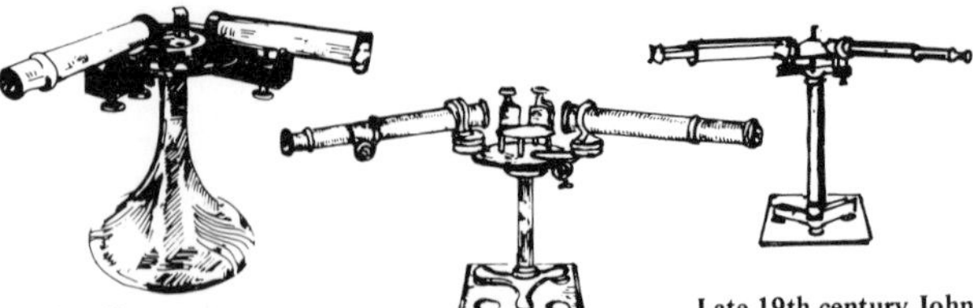

Very fine spectroscope, circa 1925, 12in. high. $169 £75

Late 19th century John Browning brass spectroscope, 12in. high, in wooden carrying case. $295 £130

Late 19th century John Browning brass spectroscope, 11in. high. $520 £230

SPHERES

German, brass, earth-centred armillary sphere with rings for the sun, moon and one other planet, the ecliptic engraved with signs of the Zodiac. $2,475 £1,100

A superb 19th century brass armillary sphere with engraved rings and standing on four turned cast brass legs and engraved base, 13in. high, 10in. diam. $3,375 £1,500

A fine, late 17th century, brass ptolemaic armillary sphere, 21ins. high. $10,690 £4,750

SPITJACKS

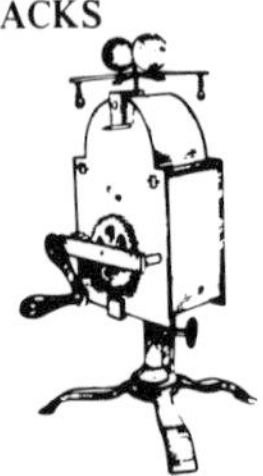

Mid 19th century French mechanical roasting jack in brass and sheet iron. $125 £55

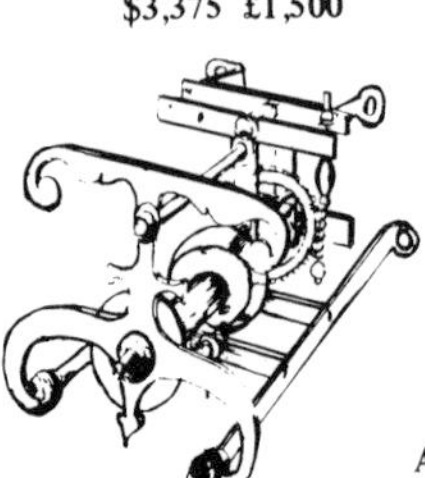

Rare, polished steel spitjack, circa 1700. $845 £375

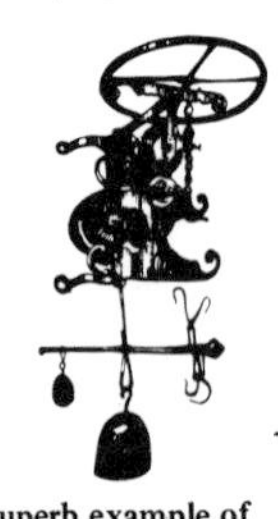

A superb example of William and Mary period, wrought iron, meat spitjack, 12in. high, circa 1690. $955 £425

Victorian chromatic stereo-scope in an inlaid walnut case. $55 £25

Victorian burr walnut folding stereoscope. $80 £35

American stereoscope made of sheet zinc and birchwood, circa 1904, 13in. long. $145 £65

Mahogany cased stereo-scope with ebonised side handles, sold with numer-ous views. $145 £65

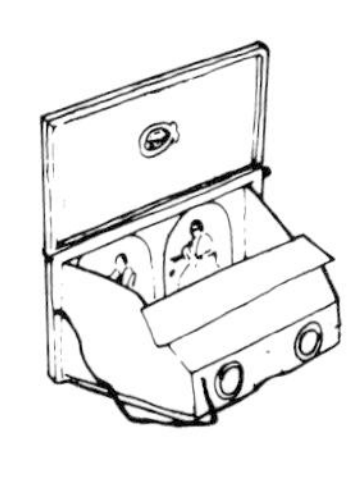

A stereoscope by T.R. Williams with Daguer-reotype portrait. $260 £115

A Victorian chromatic stereoscope on an adjustable stand. $295 £130

Ornate 19th century carved wood stereoscope viewing cabinet. $340 £150

Kilburn stereoscopic Daguerreotype portrait in collapsible viewing case, circa 1855. $350 £155

Early 20th century French 'Le Taxiphote' stereoscopic viewer, 1ft.7½in. high. $350 £155

Beck tropical box form stereoscope. $385 £170

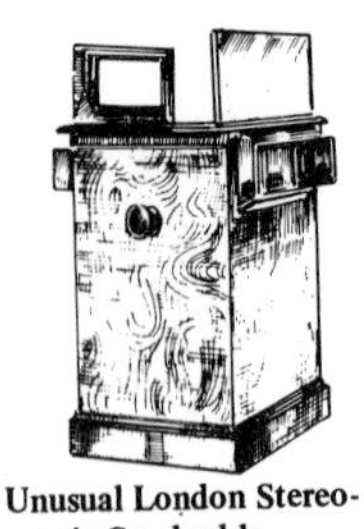

Unusual London Stereoscopic Co. double stereoscopic viewer, circa 1880, 1ft.8in. high. $395 £175

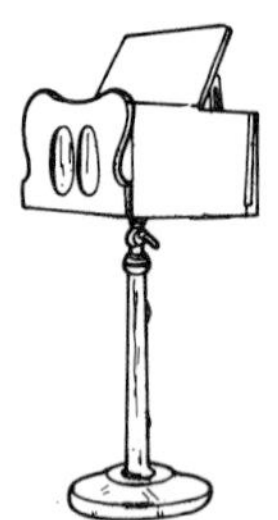

A good, table stereoscope in fitted wooden case, 15ins. high, circa 1860's. $430 £190

Negretti and Zambra Scott's patent stereoscope, circa 1880, 1ft.4½in. high. $475 £210

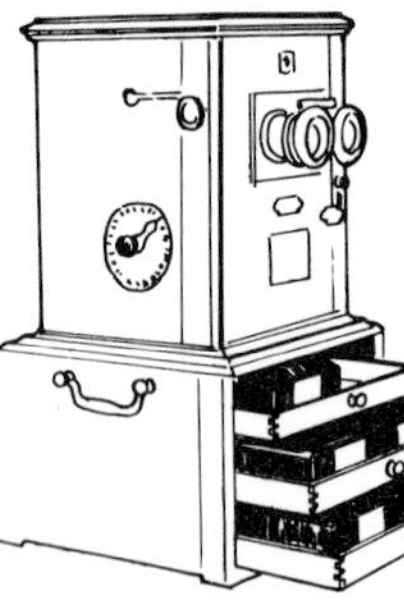

Early 19th century French 'Taxiphote' stereoscopic viewer, 1ft.7ins high. $475 £210

Victorian upright stereoscope, 121cm. high, in walnut pedestal case. $495 £220

19th century walnut stereoscopic cabinet, 18ins. tall. $595 £265

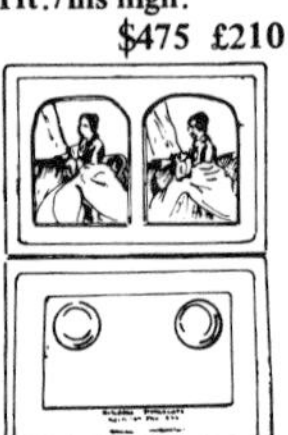

Small Kilburn stereoscopic Daguerreotype portrait, with fitted viewing case. $900 £400

Burr walnut, table model, pedestal stereoscope. $2,025 £900

A fine bronze sundial with circular base plate engraved with the points of the compass, hours and minutes, signed by Nairn and Blunt, London 1809, 10ins. diam. $170 £75

Early 19th century bronze sundial of horizontal pedestal type. $235 £105

An exceptionally fine horizontal sundial in engraved bronze, the plain gnomen with original strengthening bars either side, inscribed 'Cole Fecit', 10ins. diam. $350 £155

An exceptionally rare carved sandstone vertical wall sundial, 18in. high, hours 6am to 6pm, incised Roman numerals. $765 £340

Circular brass sundial by John Rowley, circa 1720, 30.5cm. diam. $820 £350

17th century brass sundial by Henry Wynne, London, 16¾in. diam. $1,125 £500

A bronze armillary sundial with stone pedestal, 5ft.10ins. high, 2ft.2ins. wide. $1,125 £500

A fine universal sundial by Watkins and Hill. $1,125 £500

16th century polyhedral sundial and clock in gilt metal, 5in. high. $16,650 £7,400

Boxwood cased pocket sundial and compass, circa 1860, 2in. diam. $145 £65

An early 19th century French enclosed pocket sundial in a brass case, 2¾ins. diam. $340 £150

An unusual rectangular silver pocket sundial signed Lange de Bourbon, 2½in. $700 £310

Early 17th century English brass pocket sundial signed 'John Hargrave Dyal 1638', 1½in. diam. $730 £325

German silver and gilt metal horizontal dial, 49mm. diam. $745 £330

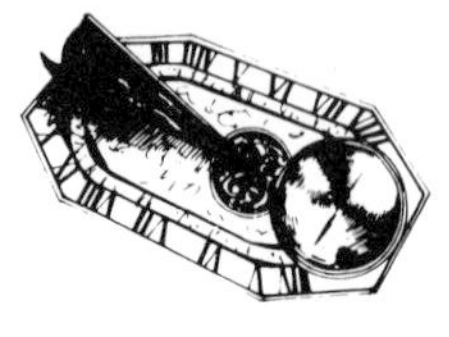

Butterfield silver dial by Febure of Paris, 2½in. $900 £400

Pocket sundial and compass combined by Michael Butterfield of Paris, early 18th century, 2¼ins. wide. $970 £430

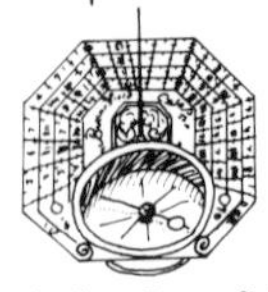

French silver Butterfield dial, signed Butterfield A Paris, fitted shagreen case, 5.3cm. greater dimension. $1,195 £500

Early 18th century silver Butterfield dial by Macquart, Paris, 65mm. long, signed. $1,195 £520

A good Butterfield type dial of silver. $1,240 £550

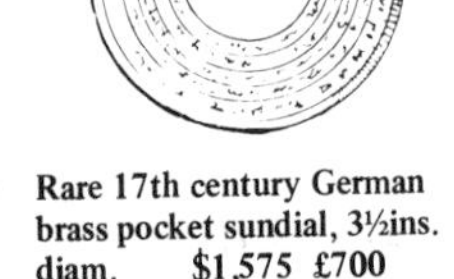

Rare 17th century German brass pocket sundial, 3½ins. diam. $1,575 £700

French silver Butterfield type dial by Chapotot, Paris, in leather case, 7cm. greater dimension. $3,110 £1,300

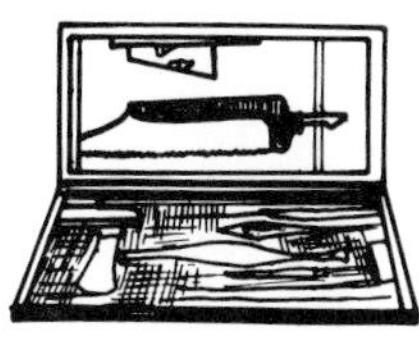

Victorian field-surgeon's kit by Plate & Co. of Manchester. $350 £155

Mid 19th century Evans & Wormull veterinary surgeon's instrument set, 1ft.6in. wide. $550 £245

Set of 19th century surgical instruments in brass bound mahogany case, 43cm. long. $810 £360

A mahogany, brass-banded, military surgeon's instrument set, circa 1840, the instruments are ebony handled and in mint condition, by J. & W. Wood, Manchester. $865 £385

Fine Crimean War set of surgeon's instruments, in original mahogany case, by Sumner & Co., Liverpool, 16¾in. x 7½in. x 3¼in., circa 1855. $990 £440

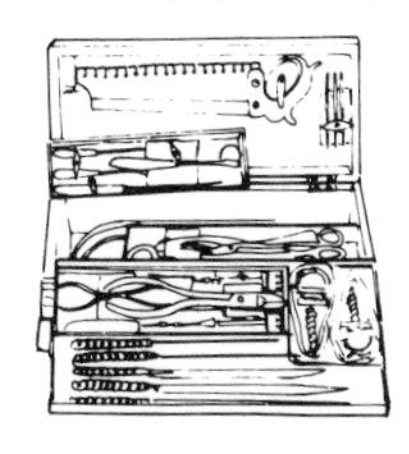

One of two brass bound mahogany cases of medical instruments, late 19th century, 43cm. wide. $1,385 £615

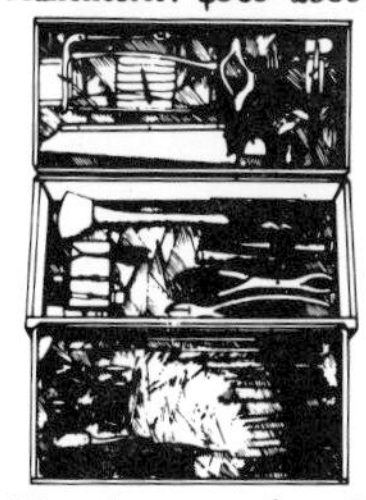

Victorian surgeon's tools contained in fitted, brass mounted, oak case containing various scalpels, tweezers, pliers, saws and mallet. $1,400 £625

18th century mahogany cased set of surgical instruments with silver fittings. $1,640 £730

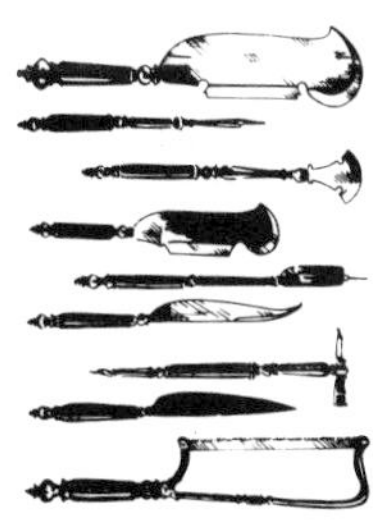

Set of South German surgical instruments, one piece dated 1569. $4,500 £2,000

SURVEYING INSTRUMENTS

Unusual surveyor's brass clinometer, circa 1870, 3in. diam. $135 £60

Surveyor's brass cylindrical cross staff in walnut case. $200 £90

Brass surveying instrument by George Beck, late 18th century, 150mm. radius. $1,410 £625

TELEGRAPHS

Late 19th century American J. H. Bunnell & Co., brass recording telegraph. $125 £55

A Hughes type-printing telegraph with an extraordinary piano keyboard. $235 £105

Late 19th century English Walters brass recording telegraph. $295 £130

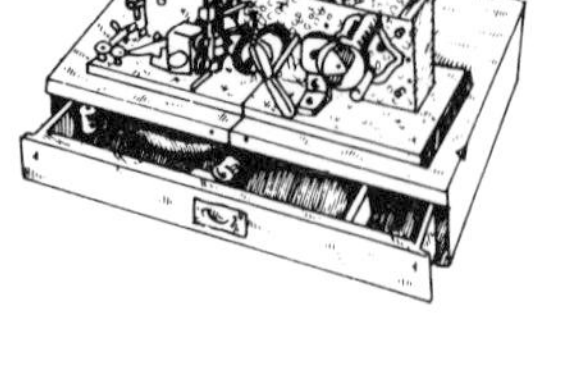

A telegraphic inkwriter by Dent from the mid 19th century. $935 £415

A British Ericsson telephone, in a walnut wall cabinet. $145 £65

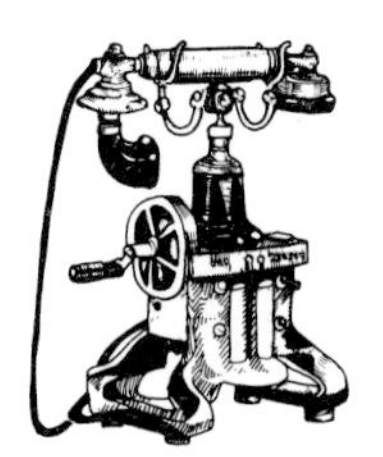

A German military field telephone, circa 1917. $180 £80

A Thomson Houston brass and copper table telephone, circa 1925. $180 £80

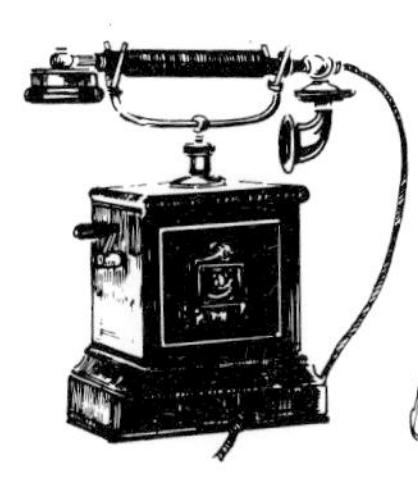

Danish magneto desk telephone, circa 1920, 1ft.1in. high. $270 £120

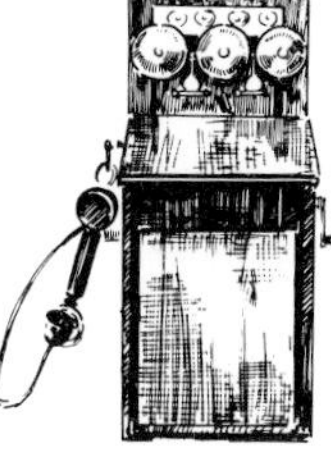

Emil Moller's magneto wall telephone, circa 1905, 2ft.3in. high. $270 £120

Ericsson type table telephone, circa 1900, 12in. high. $325 £145

Brass candlestick telephone with unusual keyboard dialling unit, circa 1925. $375 £170

An Ericsson table telephone, Swedish, circa 1895, 11in. high. $450 £200

L. M. Ericsson magneto wall telephone, circa 1900, 2ft. 4in. high. $540 £240

TELESCOPES
HAND HELD

Single draw telescope by Ross of London, 17in. long. $135 £60

Hand held telescope by Hammersley of London, 17½in. when closed, 24in. extended, circa 1840. $180 £80

19th century brass four-draw telescope with mahogany outer tube. $155 £85

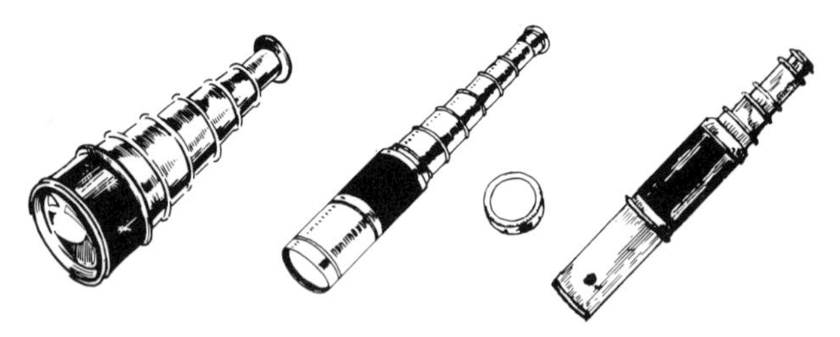

19th century six-draw monocular telescope by I. Abraham, Bath, outer tube inlaid with mother-of-pearl. $155 £85

19th century brass six-draw telescope, mahogany outer tube and dust covers. $215 £90

Brass and mahogany three draw telescope by Dolland, circa 1830. $215 £95

Naval officer's, mahogany and brass spy glass by Woodward, Clement's Inn, Strand, in original leather case. $215 £95

A fine 19th century spy glass by G. Adams, in a horn and ivory case. $540 £240

Early 18th century three-draw telescope with tooled board tubes, ebony mounts, 20.3cm. long. $760 £320

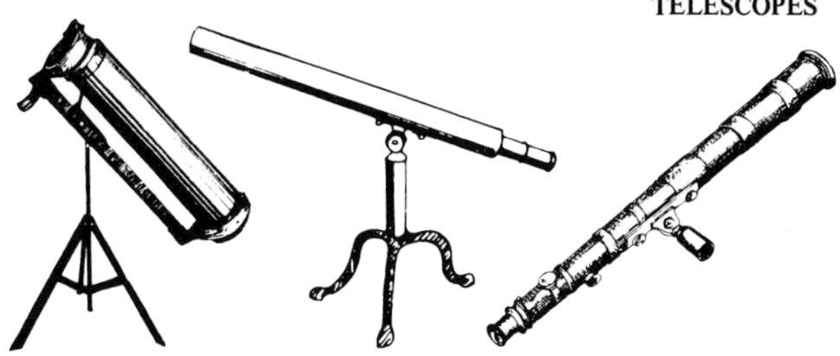

Victorian brass telescope by John Guletti, Glasgow, 10½in. long. $115 £50

Regency period brass telescope on a tripod stand. $295 £130

A Victorian mahogany and brass telescope. $350 £155

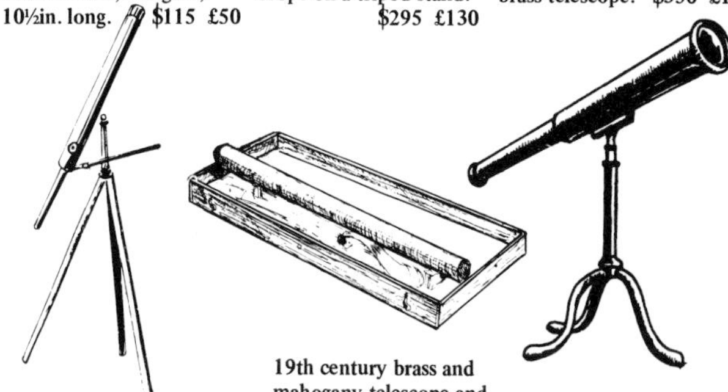

Early 19th century brass terrestrial and solar telescope. $430 £190

19th century brass and mahogany telescope and stand in a mahogany case, by Adams of London. $475 £210

Regency brass telescope on folding tripod base. $475 £210

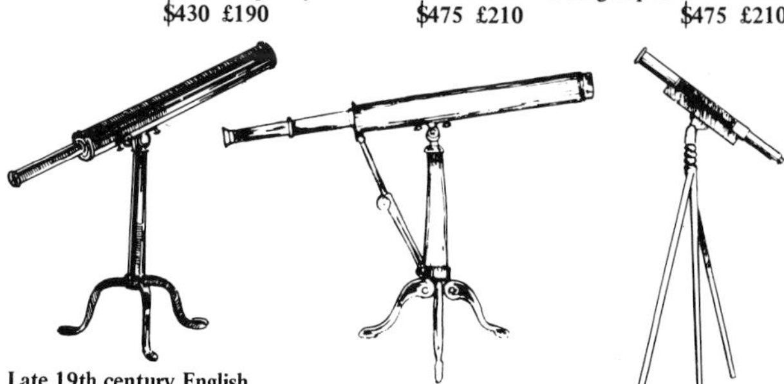

Late 19th century English brass refracting telescope on stand, tube 2ft.9in. long. $475 £210

Early 19th century table telescope by Cary of London. $495 £220

19th century brass telescope and stand. $520 £230

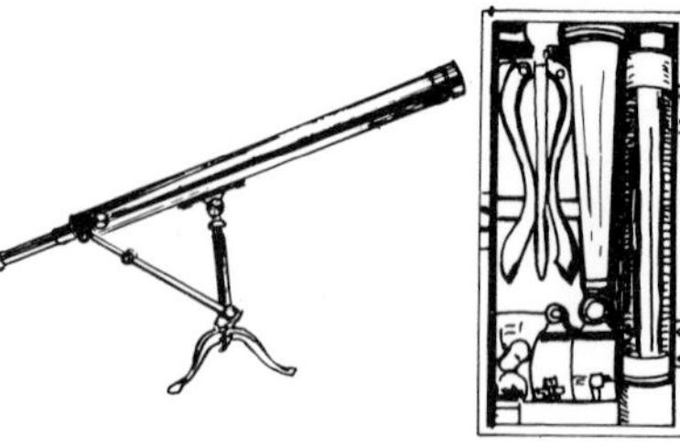

A Victorian astronomical and terrestrial telescope by Casartelli of Manchester, 39in. long. $540 £240

Brass mahogany bound three draw field telescope and stand, 19th century. $540 £240

Library telescope by Dolland of London, with a mahogany barrel, 38in. long. $585 £260

Astronomical telescope engraved H. Hughes & Sons, London, in its original pine box. $585 £260

Brass pocket telescope and stand, early 19th century, barrel diam. 39.5mm. $585 £260

19th century Georgian brass telescope. $630 £280

Early 19th century brass telescope, with folding tripod stand and original pine case, by D. McGregor & Co., Glasgow and Greenock. $700 £310

A brass astronomical telescope on folding tripod table stand, barrel engraved Troughton & Simms, London, with mahogany box. $700 £310

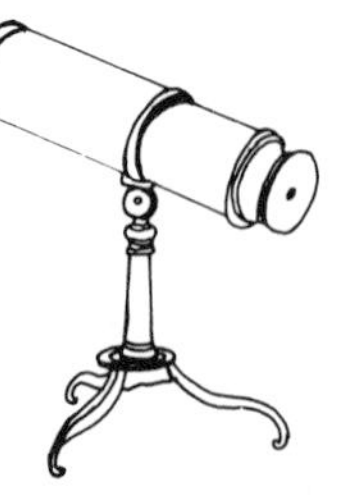

Mid 19th century refracting telescope by W. & S. Jones, London. $730 £325

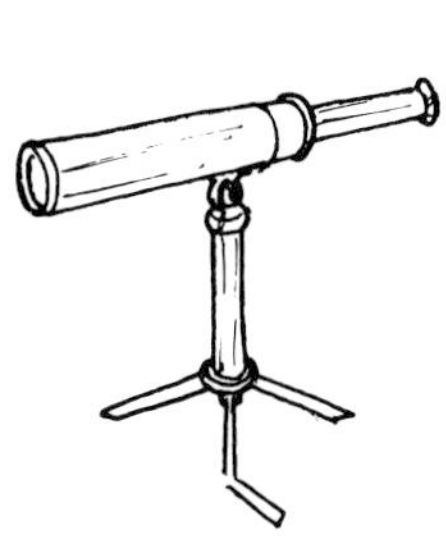

Brass single draw telescope and stand by John Noiland, circa 1752. $730 £325

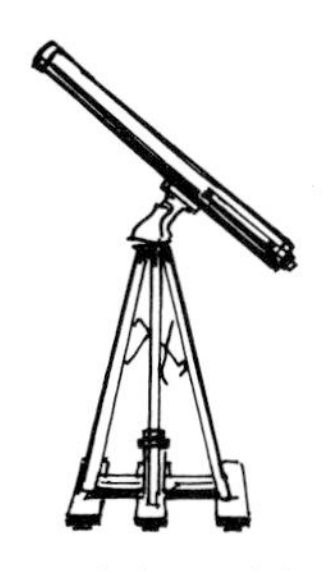

Late 19th century 3½in. brass refracting astronomical telescope, 4ft.1in. long. $765 £340

Late 19th century 4in. brass refracting observatory telescope by T. Cook & Sons Ltd., 5ft.4½in. long. $775 £345

Late 19th century brass refracting telescope. $775 £345

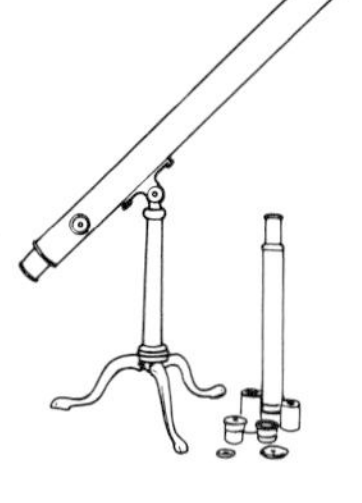

Early 19th century 2in. Gilbert & Co. refracting telescope on stand. $820 £365

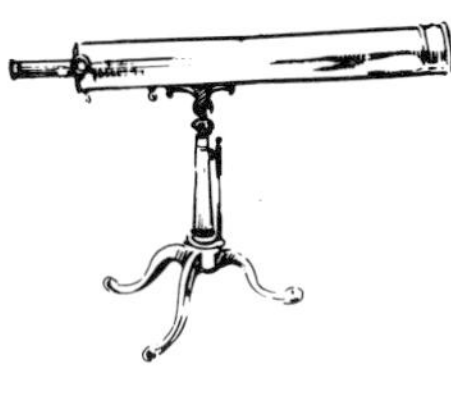

Brass astronomical telescope by Harris. $845 £375

Late 19th century Dolland brass refracting telescope on stand, tube 3ft.5in. long. $865 £385

An 18th century oxidised brass reflecting telescope signed S. Johnson, London. $900 £400

Mid 19th century Wm. Harris brass refracting telescope on stand, tube 2ft. 6in. high. $945 £420

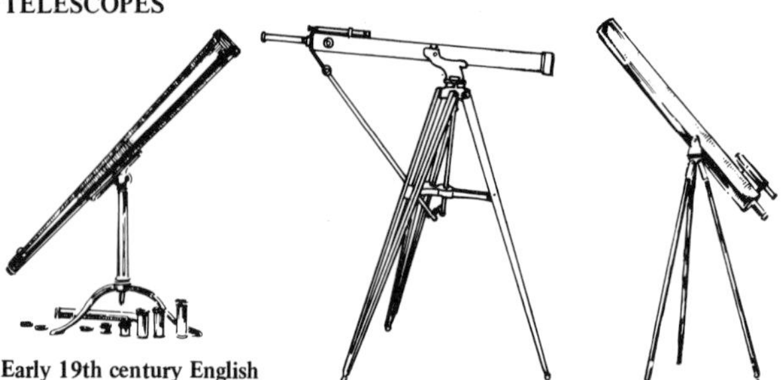

Early 19th century English Ramsden brass refracting telescope on stand, tube 3ft. long. $1,125 £500

Late 19th century, 4-inch, Callaghan & Co., refracting telescope. $1,125 £500

Brass celestial telescope on mahogany tripod stand, 45in. long. $1,200 £535

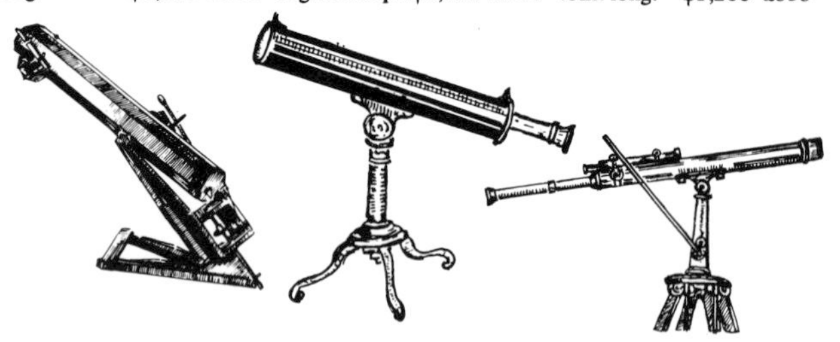

Late 19th century 'Herschel type' Newtonian refracting telescope, tube 4ft.4in. long. $1,240 £550

Brass Gregorian telescope by J. Dolland & Son, London, late 18th century. $1,240 £550

19th century brass cased astronomical telescope by J. Lancaster & Sons Ltd., Birmingham. $1,350 £600

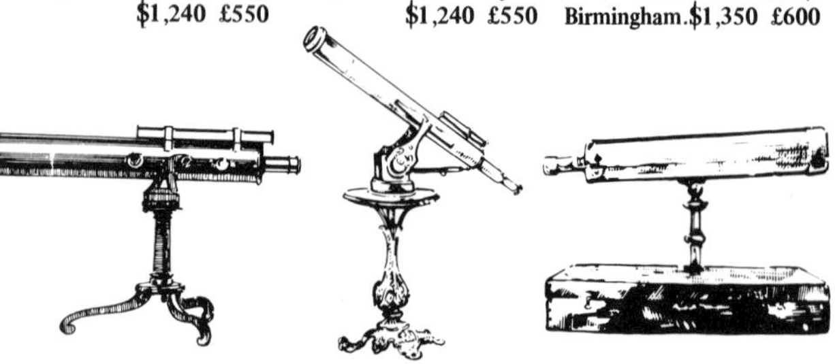

Early 19th century Wm. Struthers Gregorian reflecting telescope on stand, tube 2ft.0½in. long. $1,360 £605

Mahogany cased brass reflecting telescope, circa 1760, 19½in. long. $1,610 £715

19th century brass astronomical telescope on a steel stand by Jas. Parker & Son. $1,585 £705

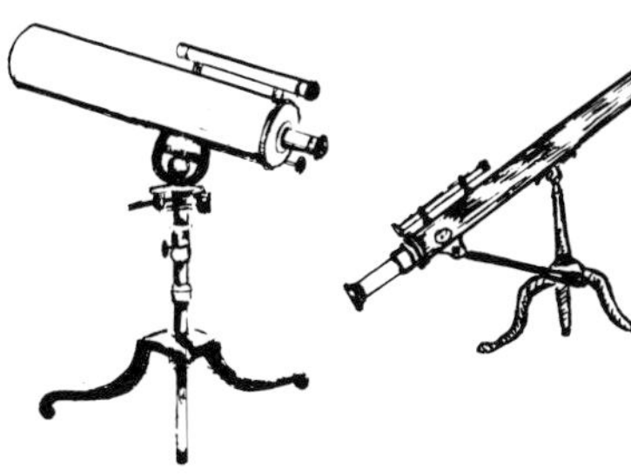

An 18th century brass Georgian type reflecting telescope by Jas. Short of London, 24¼in. long. $1,755 £780

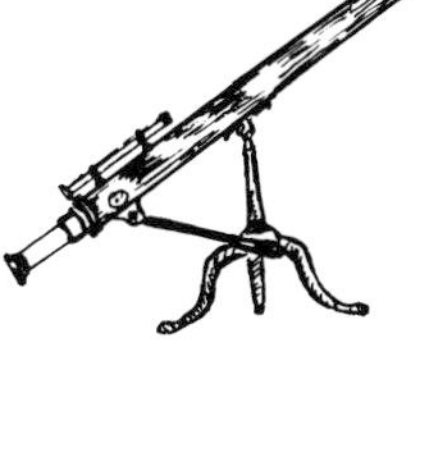

Astronomical brass telescope by J. H. Steward, London. $1,825 £810

Mahogany lacquered brass achromatic telescope, 168cm. long. $1,855 £825

Troughton & Simms refracting telescope on stand, 1857, tube 3ft.7in. long. $1,855 £825

Rare mid 19th century transit telescope, 16in. high, by Troughton & Simms. $2,025 £900

Superb bronzed brass 4in. telescope on a mahogany tripod by Thomas Cooke of York. $2,080 £925

Large brass telescope by W. S. Jones, London, circa 1800. $2,250 £1,000

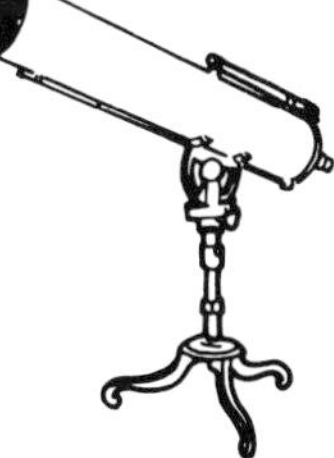

A fine quality brass Gregorian reflecting telescope by James Short, 1763, 24in. long. $2,925 £1,300

Early 19th century Herschel 7ft. Newtonian reflecting telescope. $16,310 £7,250

Brass theodolite, signed Cooke, Troughton & Simms, in fitted case, 29.8cm. high. $210 £90

Mid 19th century American brass Altizimuth theodolite by N. & L. E. Gurley, 1ft. high. $350 £155

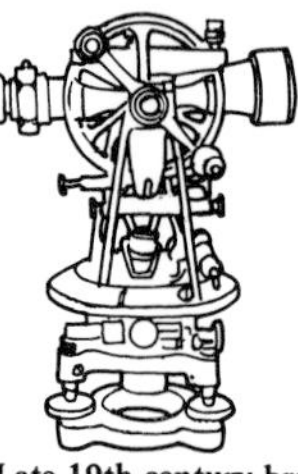

Late 19th century brass Altizimuth theodolite by Cooke, Troughton & Simms, 1ft. high. $395 £175

Early 20th century Troughton & Simms oxidised brass transit theodolite, 15in. high. $525 £240

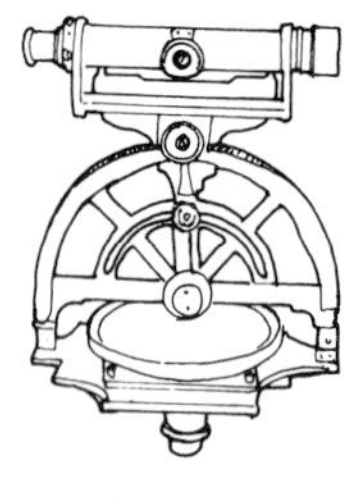

Late 19th century Troughton & Simms brass Altizimuth theodolite in a mahogany case. $640 £285

Early 20th century French oxidised brass transit theodolite by Breithaupt & Sohn, 6¾in. high. $650 £300

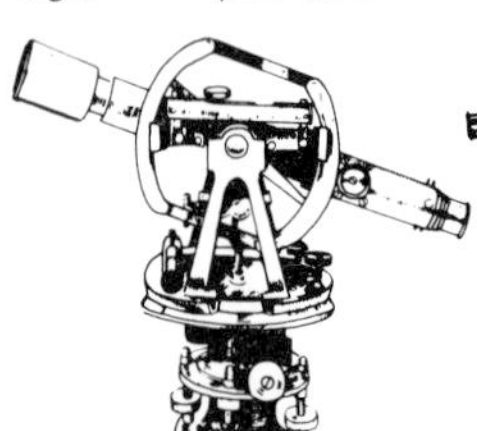

Mid 19th century Casella brass transit theodolite, 20cm. high, in mahogany case. $835 £350

Early 20th century J. B. Winter brass transit theodolite, 1ft.1½in. high. $745 £330

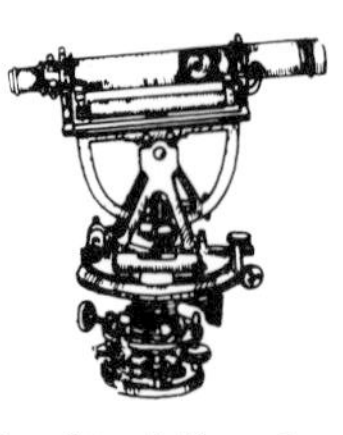

Troughton & Simms brass Y-type theodolite, circa 1880, 28cm. high, in mahogany case. $905 £380

Two scope brass theodolite by Schmalcalder of London. $980 £435

Late 19th century Troughton & Simms brass transit theodolite, 1ft.2in. high. $1,035 £460

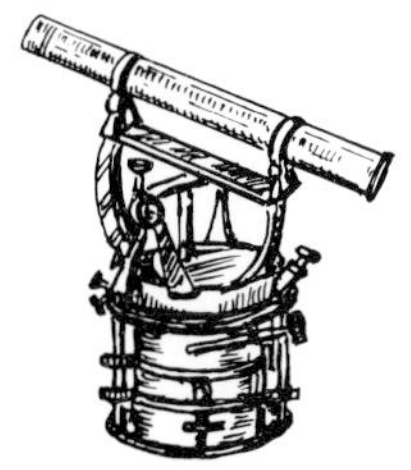

Surveyor's brass theodolite by J. Davis, 12¾in. high. $1,090 £485

Surveyor's fine theodolite by Charles Baker, London, 13in. high. $1,105 £485

Early 19th century brass theodolite of small size, by W. & S. Jones, London. $1,215 £660

18th century brass theodolite by Adams of Charing Cross. $1,520 £675

18th century brass theodolite by Dolland of London. $1,860 £825

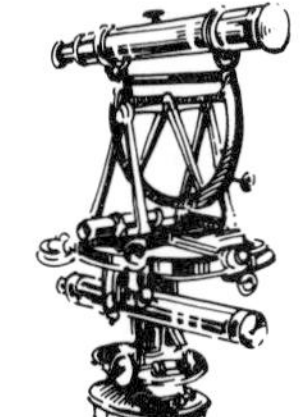

Good brass theodolite, 350mm. high. $2,025 £900

Mid 18th century brass theodolite by T. Heath, London. $2,925 £1,300

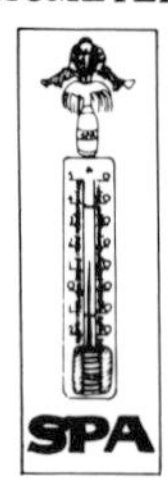

Enamelled advertising sign for Spa by Email Belg, 27 x 8½in. $125 £55

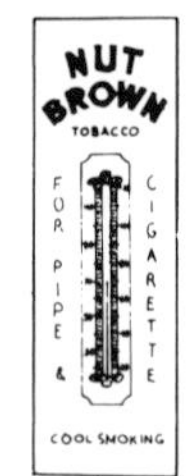

Enamelled advertising sign for Nut Brown Tobacco in black and white on a red ground, 22½ x 7½in. $95 £40

A thermometer in a folding ivory case. $115 £50

George III wall hanging thermometer by Lione & Co., Hatton Garden, London, circa 1820, 18in. long. $340 £150

Victorian barometer, clock and thermometer in oak case, 13in. high. $215 £95

Unusual advertising sign for Stephen's Ink, by Jordan-Belston, with domed top 60 x 12in. $405 £180

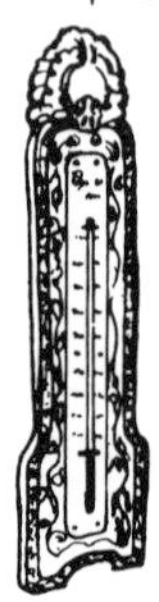

Early 19th century French desk or wall thermometer. $415 £185

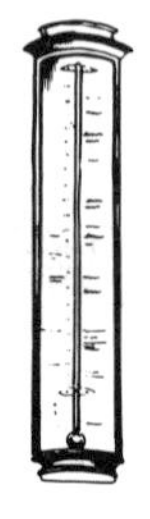

Unusual bow front thermometer by Dolland. $1,350 £620

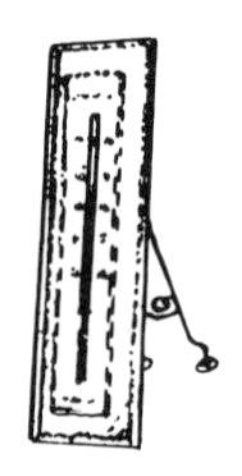

Carl Faberge desk thermometer, workmaster K. Armfelt, 7¼ x 2¼in. $4,500 £2,000

Late 19th century silver backed manicure set in a fitted case. $55 £25

Victorian mahogany vanity case with silver plated fittings, circa 1860. $170 £75

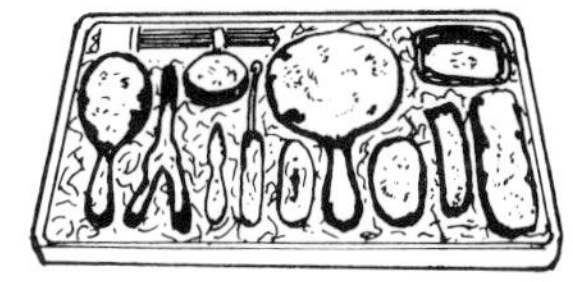

Turtle shell toilet set with Chinese dragon design, circa 1900-1910. $300 £135

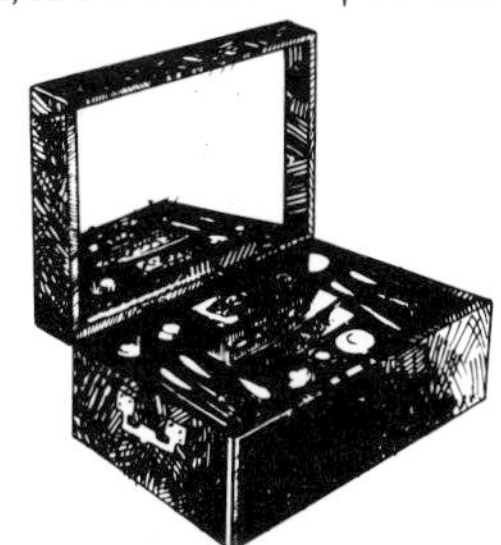

Fine 19th century dressing case, London, 1838, with engraved silver fittings. $860 £385

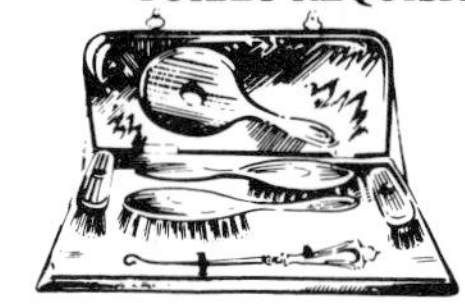

Seven-piece dressing table set with silver mounts. $70 £30

Ebony oval etui case with recesses for scissors, etc., in French hallmarked gold and silver. $260 £115

Early Victorian dressing case with nine engraved silver fittings, 1841, 11in. wide. $720 £320

Portuguese silver and stained fishskin necessaire de voyage, circa 1730-40. $1,290 £575

Later keyboard type model of a typewriter. $80 £35

American Simplex Model D typewriter in original cardboard box, 1930's, 8¾in. wide. $85 £40

Mid 19th century Edwin Ponting multiple writing machine, 17in. wide. $170 £75

English Imperial Model B typewriter, circa 1915, 1ft. wide. $190 £85

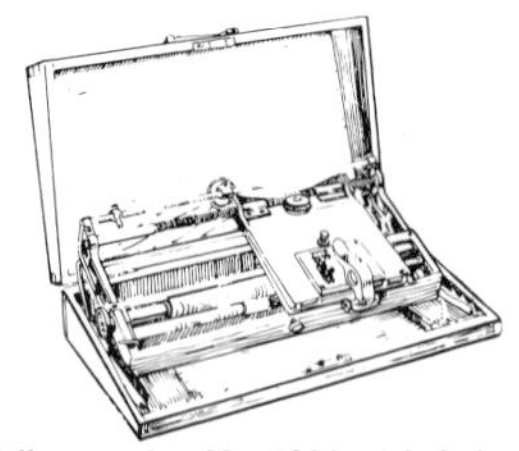

Hall typewriter No. 5380, nickel plated in mahogany case, circa 1887, 15½in. wide. $390 £180

Good, American Lambert typewriter, circa 1900. $470 £210

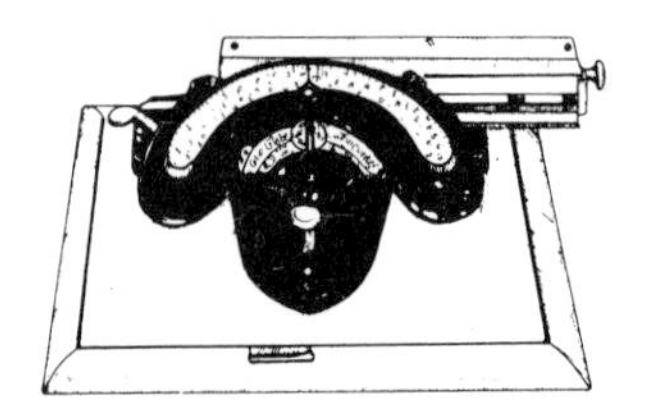

American Globe typewriter by the American Typewriter Co., circa 1895. $510 £220

A rare 'Invincible' typewriter, No. 498, 12in. long. $525 £235

Unusual Odell typewriter No. 4, circa 1890, 10in. square. $540 £240

Mid 1890's Salter typewriter, in good condition. $585 £260

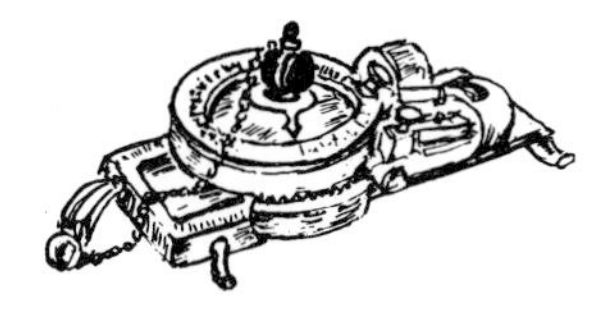

Unusual 'Virotyp' portable typewriter, French, circa 1914, 6¼in. long. $675 £300

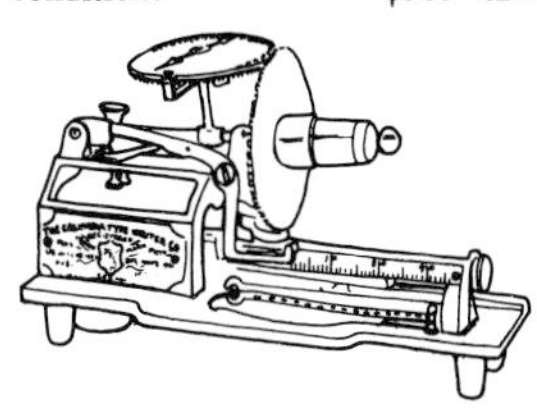

Good American Columbia typewriter, 11in. wide, circa 1890. $820 £365

Rare Victor typewriter, American, circa 1895, 12in. wide. $1,485 £660

Early model of a typewriter by Columbia. $2,140 £950

Early Swiss 'Velograph' typewriter, circa 1887. $2,475 £1,100

Early Shokes & Gliddon Remington typewriter. $5,235 £2,200

WATCHMAKER'S INSTRUMENTS

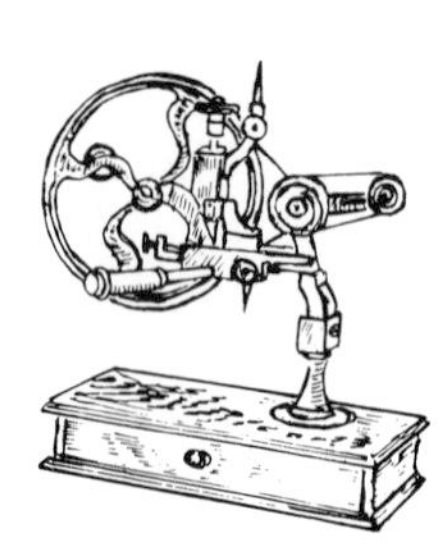

Watchmaker's brass topping tool, 12in. long. $415 £185

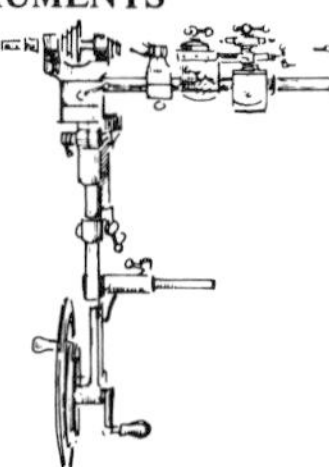

Early 20th century F. Lorch's watchmaker's lathe, 14¼in. long, sold with twenty tools. $620 £275

19th century watchmaker's Swiss brass mandrel, 420mm. long. $675 £300

WAYWISERS

Mahogany waywiser by Fraser, Bond St., London, recording the distance measured in miles, furlongs, poles and yards. $640 £285

Early 19th century George Adams waywiser, 4ft.4in. high, with cast iron handles. $745 £330

Rare 18th century waywiser with 23in. diam. steel wheel and handles of turned mahogany, made by B. Cole, Fleet Street, London. $1,240 £550

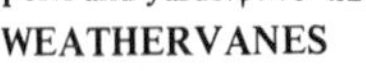

WEATHERVANES

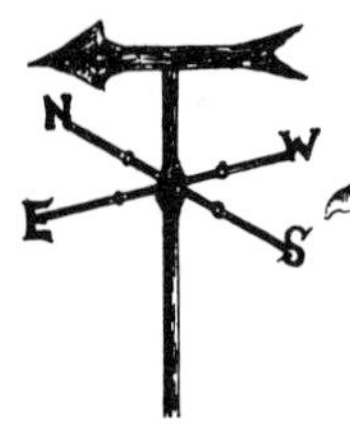

A Victorian copper weathervane, 3ft.1in. high. $115 £50

American moulded metal Gabriel weathervane, 31½in. wide, circa 1870. $500 £230

19th century iron weathervane with copper ball, overall height 6ft.3in. $500 £230

WEATHERVANES

A metal weathervane of a flying griffon, the Roman capital letters in thick copper, circa 1810. $585 £260

Decorative 18th century copper and wrought iron weathervane, the whole surmounted by an imposing silhouette wolf of polished steel, overall height 65in. $630 £280

Rare 17th century copper weathervane in the shape of the 'Mayflower'. $700 £310

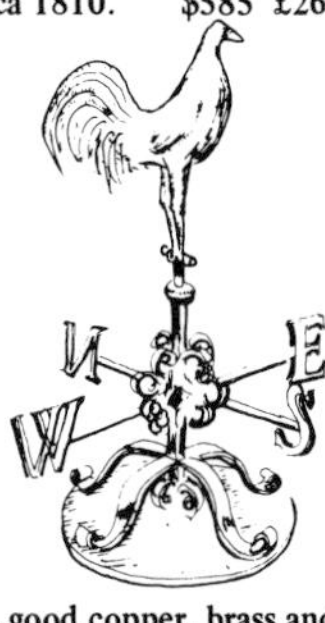

A good copper, brass and wrought iron weather cock, 40in. high. $935 £415

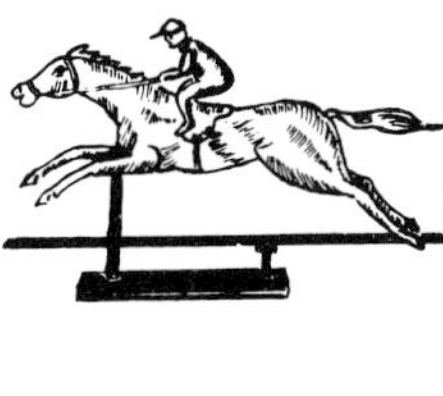

American copper horse and rider weathervane, circa 1900, 31in. long. $1,500 £650

Mid 19th century American metal rooster weathervane, 24in. high. $1,500 £695

ZEOTROPES

English zeotrope, tin drum on carved wood stand, circa 1870, 12in. diam. $300 £130

Late 19th century zeotrope, tin on wood, 14in. high. $525 £240

Mid 19th century English zeotrope, 14in. high. $845 £375

INDEX